ORBITAL MECHANICS

ORBITAL MECHANICS

JOHN E. PRUSSING

Professor of Aeronautical and Astronautical Engineering
University of Illinois at Urbana-Champaign

BRUCE A. CONWAY

Professor of Aeronautical and Astronautical Engineering
University of Illinois at Urbana-Champaign

New York   Oxford
Oxford University Press
1993

Oxford University Press

Oxford New York Toronto
Delhi Bombay Calcutta Madras Karachi
Kuala Lampur Singapore Hong Kong Tokyo
Nairobi Dar es Salaam Cape Town
Melbourne Auckland Madrid

and associated companies in
Berlin Ibadan

Copyright © 1993 by Oxford University Press, Inc.

Published by Oxford University Press, Inc.,
200 Madison Avenue, New York, New York 10016

Oxford is a registered trademark of Oxford University Press

All rights reserved. No part of this publication
may be reproduced, stored in a retrieval system, or transmitted,
in any form or by any means, electronic, mechanical,
photocopying, recording, or otherwise, without the prior
permission of Oxford University Press.

Library of Congress Cataloging-in-Publication Data
Prussing, John E.
Orbital mechanics / John E. Prussing, Bruce A. Conway
p. cm. Includes bibliographical references and index.
ISBN 0-19-507834-9
1. Orbital mechanics. I. Conway, Bruce A. II. Title.
TL 1050.P78 1993 629.4′1133—dc20 92–41505

Printing (last digit): 9 8 7 6 5 4 3 2 1

Printed in the United States of America
on acid-free paper

For Laurel, Heidi, Erica, and Nickie

For Linda and Charles Alvin

Preface

This text takes its title from an elective course at the University of Illinois at Urbana-Champaign that has been taught to senior undergraduates and first-year graduate students for the past 22 years. Many of these students chose aerospace engineering because of their keen interest in space exploration. For them, the senior-year elective courses, such as orbital mechanics, rocket propulsion, and spacecraft design, are the reason they came to the university.

We attempt to develop the subject of orbital mechanics starting from the first principles of Newton's Laws of Motion and the Law of Gravitation. While it is not unusual in an introductory book to derive Kepler's Laws of Planetary Motion from Newton's laws, we also derive the other important results: Lambert's equation, the rocket equation, the hyperbolic gravity-assist relations, the Hill–Clohessy–Wiltshire equations of relative motion, the Lagrange perturbation equations, and the Gauss and Laplace methods of orbit determination, from first principles.

There is more material in the text than we can present in one semester. We customarily cover Chaps. 1 through 7, after which the student is well-versed in the basic fundamentals and ready to study advanced topics. Orbit transfer receives special emphasis because it is a favorite research area of the authors and their graduate students. There is usually time remaining to cover at least one of the remaining three chapters: Chap. 8, on linear orbit theory, which is important to problems of spacecraft rendezvous, Chap. 9, on the effect of perturbations such as atmospheric drag and earth oblateness, and Chap. 10, on orbit determination from observations. All of these subjects are important, even for an introduction to orbital mechanics, but each instructor can choose which topics to emphasize.

It is customary in the last paragraph of a preface for the authors to thank those who did the typing, the graphics, and the proofreading. With computer word processing and drawing programs, it is virtually as easy to do such things yourself as supervise someone else's work, so we have only ourselves to thank or blame. The typesetting was done using the *troff* package under the Unix operating system, and the graphics were done on a Macintosh computer using the SuperPaint program. We do, however, wish to thank the many years worth of AAE 306 students who provided the inspiration to write this book, those who brought errors in previous versions to our attention over the past six years, Dan Snow and Denise Kaya who gave us valuable information on how orbit determination is actually done, and several anonymous reviewers whose suggestions were helpful.

J. E. P.
B. A. C.
Urbana, Ill.
December 1992

Contents

1 The n - Body Problem, 3
    1.1 Introduction, 3
    1.2 Equations of Motion for the n - Body Problem, 6
    1.3 Justification of the Two-Body Model, 9
    1.4 The Two-Body Problem, 12
    1.5 The Elliptic Orbit, 15
    1.6 Parabolic, Hyperbolic, and Rectilinear Orbits, 17
    1.7 Energy of the Orbit, 19
        References, 22
        Problems, 22

2 Position in Orbit as a Function of Time, 26
    2.1 Introduction, 26
    2.2 Position and Time in an Elliptic Orbit, 26
    2.3 Solution for the Eccentric Anomaly, 30
    2.4 The f and g Functions and Series, 32
    2.5 Position versus Time in Hyperbolic and Parabolic Orbits: Universal Variables, 36
        References, 42
        Problems, 42

3 The Orbit in Space, 46
    3.1 Introduction, 46
    3.2 The Orbital Elements, 46
    3.3 Determining the Orbital Elements from r and v, 49
    3.4 Velocity Hodographs, 54
        Reference, 59
        Problems, 59

4 Lambert's Problem, 62
    4.1 Introduction, 62
    4.2 Transfer Orbits Between Specified Points, 62
    4.3 Lambert's Theorem, 67
    4.4 Properties of the Solutions to Lambert's Equation, 70
    4.5 The Terminal Velocity Vectors, 75
    4.6 Applications of Lambert's Equation, 78
        References, 79

Problems, 79

5 Rocket Dynamics, 81
- 5.1 Introduction, 81
- 5.2 The Rocket Equation, 81
- 5.3 Solution of the Rocket Equation in Field-Free Space, 83
- 5.4 Solution of the Rocket Equation with External Forces, 87
- 5.5 Rocket Payloads and Staging, 88
- 5.6 Optimal Staging, 92
  - References, 97
  - Problems, 97

6 Impulsive Orbit Transfer, 99
- 6.1 Introduction, 99
- 6.2 The Impulsive Thrust Approximation, 99
- 6.3 Two-Impulse Transfer between Circular Orbits, 102
- 6.4 The Hohmann Transfer, 104
- 6.5 Coplanar Extensions of the Hohmann Transfer, 108
- 6.6 Noncoplanar Extensions of the Hohmann Transfer, 112
- 6.7 Conditions for Intercept and Rendezvous, 114
  - References, 117
  - Problems, 118

7 Interplanetary Mission Analysis, 120
- 7.1 Introduction, 120
- 7.2 Sphere of Influence, 121
- 7.3 Patched Conic Method, 124
- 7.4 Velocity Change from Circular to Hyperbolic Orbit, 128
- 7.5 Planetary Flyby (Gravity-Assist) Trajectories, 129
- 7.6 Flyby Following a Hohmann Transfer, 134
  - References, 137
  - Problems, 137

8 Linear Orbit Theory, 139
- 8.1 Introduction, 139
- 8.2 Linearization of the Equations of Motion, 139
- 8.3 The Hill–Clohessy–Wiltshire (CW) Equations, 142
- 8.4 The Solution of the CW Equations, 144
- 8.5 Linear Impulsive Rendezvous, 150
  - References, 153
  - Problems, 153

Contents xi

9 Perturbation, 155
    9.1 Introduction, 155
    9.2 The Perturbation Equations, 155
    9.3 Effect of Atmospheric Drag, 164
    9.4 Effect of Earth Oblateness, 164
        References, 168
        Problems, 168

10 Orbit Determination, 170
    10.1 Introduction, 170
    10.2 Angles-Only Orbit Determination, 172
    10.3 Laplacian Initial Orbit Determination, 173
    10.4 Gaussian Initial Orbit Determination, 176
    10.5 Orbit Determination from Two Position Vectors, 180
    10.6 Differential Correction, 181
        References, 185
        Problems, 186

Appendix 1 Astronomical Constants, 188

Appendix 2 Physical Characteristics of the Planets, 189

Appendix 3 Elements of the Planetary Orbits, 190

Index, 191

ORBITAL MECHANICS

1
The *n*-Body Problem

1.1 Introduction

Space mechanics is the science concerned with the description of the motion of bodies in space. The subject is divided into two parts, with one regarding the motion of spacecraft, which is termed *astrodynamics*, and the other regarding the motion of natural bodies, which is referred to as *celestial mechanics*. The motion of a spacecraft in unpowered or coasting flight is essentially the same as that of a small, natural body, so that the basic theory we will develop applies in both cases. Astrodynamics is commonly divided into consideration of either spacecraft attitude motion *about* its center of mass or the orbital motion *of* its center of mass. The latter subject will concern us here.

Johannes Kepler provided the first quantitative statements about orbital mechanics. As described in [1.1]:

> The German astronomer Johannes Kepler (1571–1630) worked briefly with the Danish astronomer Tycho Brahe, who gathered some of the most accurate observational data on planetary motion in the pre-telescope era. When Brahe died in 1601, Kepler inherited his data books and devoted many years of intensive effort to finding a mathematical description for the planetary motion described by the data. Kepler was successful in deriving three laws of planetary motion which led ultimately to our current understanding of the orbital motion of planets, moons and comets, as well as man-made satellites and spacecraft. The first two laws were published in 1609, about the time Galileo was first making astronomical observations with his telescope; the third law did not appear until a decade later, in 1619. Briefly stated, these laws are:
>
> 1. **The orbit of each planet is an ellipse with the sun at one focus.**
>
> 2. **The line from the sun to a planet sweeps out equal areas inside the ellipse in equal lengths of time.**
>
> 3. **The squares of the orbital periods of the planets are proportional to the cubes of their mean distances from the sun.**

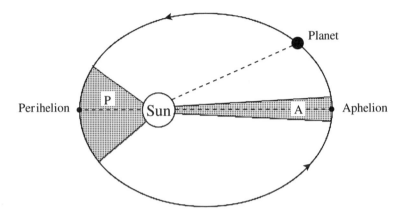

Fig. 1.1. Elliptic orbit of a planet with sun at one focus. Equal areas P and A are swept out in equal lengths of time.

The *first law* describes the geometrical shape of the orbit as an ellipse, which correctly accounts for the manner in which the distance from the sun to the planet changes as the planet travels along its orbit. As shown in Fig. 1.1, the *perihelion* is the point in the orbit at which the planet is closest to the sun; the *aphelion* is the point farthest from the sun.

Prior to Kepler's time, it was assumed that the planetary orbits were concentric circles with the sun at the center, because it was thought that only the perfect geometrical curve of the circle could describe the motion through the heavens of perfect celestial bodies. However, as the accuracy of the observational data increased, it became clear that the distance from the sun of a planet was not constant, as it would be for a circular orbit. Increasingly complicated geometrical constructions using several circles were proposed in an attempt to fit the observational data for an individual planet orbit. Kepler was the first person to recognize that an ellipse, rather than a circle, is the geometrical curve which describes the shape of a planet orbit in a simple and elegant manner.

The *second law* is a concise mathematical description of the observed fact that the rate at which the sun–planet line rotates through space (the angular velocity of the motion) increases as the planet moves closer to the sun and decreases as it moves farther from the sun. As shown in Fig. 1.1, the distance along the orbit traveled near the perihelion in given length of time (for example, one month) is greater that the distance traveled near aphelion during the same length of time. This phenomenon is described qualitatively by the statement that equal areas inside the ellipse are swept out by the sun–planet line in equal lengths of time. The time required to sweep out the total area inside the ellipse is the orbit period of the planet.

The first two laws describe the motion of an individual planet. By contrast, the *third law* states the manner in which the motions of the various planets are related to each other. It states that the ratio formed by dividing the square of the orbit period of any planet by the cube of its mean distance from the sun is the same value for all planets in our solar system. The term *mean distance*, as used in the third law, is simply the average of the perihelion and aphelion distances. This mean distance is then half of the distance between the perihelion and the aphelion, and is called the *semimajor axis* of the ellipse. Another way of stating the third law is that the period of the planet is proportional to the 3/2 power of its mean distance from the sun.

Kepler, to his credit, formulated these three laws based entirely on empirical data, without the benefit of a fundamental theory which explained why planetary motion satisfied these three laws. The missing ingredient was the concept of gravitational force, which was developed several decades later by Sir Isaac Newton. Newton, starting with Kepler's first law, deduced that a planet would move in an elliptic orbit with the sun at one focus only if the force exerted on the planet by the sun was proportional to the inverse square of the distance between them. This is the so-called *inverse-square law* of gravitational force. Newton also showed that Kepler's second law was a consequence of the principle of conservation of angular momentum. The angular momentum is conserved because the gravitational force is a central force, that is, it acts along the line from the planet to the central body, the sun. The third law is a natural consequence of the inverse-square gravitational force field of the sun, which exists throughout the solar system. It is interesting to note that the motion of any body orbiting the sun in a closed, periodic orbit is governed by the same force law and must satisfy Kepler's laws. Thus the motion of comets, asteroids, and interplanetary spacecraft during periods when the rocket thrust is off, such as the Voyager and Galileo spacecraft launched to Jupiter in 1977 and 1989, are governed by these same laws.

As we will see, the motion of natural and artificial satellites about planets are also governed by the same laws, the difference being that the central body is a planet rather than the sun.

In the course of our analytical development of the equations of orbital mechanics and their solutions, we will rediscover Kepler's laws as a natural consequence of our investigations. The basis of the analytical description of the motion of bodies in space is a combination of two of Newton's laws: the second law of motion and the law of gravitation. In the form in which it will be useful to us, the second law of motion may be expressed as:

$$\mathbf{F} = \frac{d}{dt}(m\mathbf{v}) \qquad (1.1)$$

That is, the external force applied to a body is equal to the time rate of change of the linear momentum of the body. Newton's law of gravitation may be expressed as:

$$\mathbf{F} = \frac{Gm_1 m_2}{r^2}\left[\frac{\mathbf{r}}{r}\right] \qquad (1.2)$$

That is, the force on body 1 due to attraction of body 2 depends directly on the product of their masses, inversely on the square of the separation distance r, and has the direction of the unit vector ($\mathbf{r}/r$), where $\mathbf{r}$ locates m_2 from m_1. In Eq. (1.1) the vector derivative is relative to an inertial observer. The universal constant of gravitation, G, has the value 6.67259×10^{-11} m^3 kg^{-1} s^{-2}, but it will turn out that the product of G and the mass of each major celestial body is the most useful constant for orbit calculations. This product will be cataloged for each celestial body.

1.2 Equations of Motion for the n - Body Problem

The point mass n- body problem has occupied astronomers and mathematicians for three centuries. The problem may be simply stated: Given the positions and velocities of bodies of known mass at some initial time, find the positions and velocities of the bodies at any other time. It is clearly a very practical problem; it is precisely the problem to be solved for the guidance of a spacecraft from one planet to another, subject to the attraction of the sun and all the planets in the solar system

Let the vector $\mathbf{R}_i$ represent the position of a body, whose mass is m_i, with respect to the origin O of an inertial reference frame, as shown in Fig. 1.2. The position of the jth mass with respect to the ith mass will then be called $\mathbf{r}_{ij}$, where

The n-Body Problem

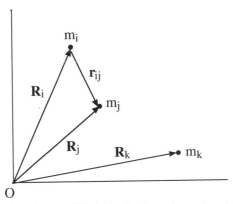

Fig. 1.2. Absolute and Relative Position in an Inertial Frame

$$\mathbf{r}_{ij} = \mathbf{R}_j - \mathbf{R}_i, \quad i = 1, 2, \cdots, n \tag{1.3}$$

The attraction of the ith body is then determined by the attraction of the n-1 other bodies for it. Using Eqs. (1.1) and (1.2) and summing over the system of masses yields:

$$m_i \ddot{\mathbf{R}}_i = \mathbf{F}_i = G \sum_{j=1}^{n} \frac{m_i m_j}{r_{ij}^3} \mathbf{r}_{ij} \quad (j \neq i) \tag{1.4}$$

Now, since $\mathbf{r}_{ij} = -\mathbf{r}_{ji}$, Eq. (1.4) may be summed over all of the bodies in the system to yield:

$$\sum_{i=1}^{n} m_i \ddot{\mathbf{R}}_i = 0 \tag{1.5}$$

which may be integrated once to yield:

$$\sum_{i=1}^{n} m_i \dot{\mathbf{R}}_i = \mathbf{C}_1, \text{ a } \mathbf{constant} \text{ vector}, \tag{1.6}$$

and integrated again to yield

$$\sum_{i=1}^{n} m_i \mathbf{R}_i = \mathbf{C}_1 t + \mathbf{C}_2 \tag{1.7}$$

Recalling the definition of center of mass, $\mathbf{R}_{cm} \equiv \sum m_i \mathbf{R}_i / \sum m_i$, Eq. (1.7) determines the motion of the system center of mass, which is rectilinear and, from Eq. (1.6) at constant velocity. The linear momentum of the system is thus conserved, which is expected since the system of bodies is subject to no net external force.

Now, taking the vector product of $\mathbf{R}_i$ with each term on the left-hand side of Eq. (1.4) yields

$$\sum_{i=1}^{n} m_i \mathbf{R}_i \times \ddot{\mathbf{R}}_i = G \sum_{i=1}^{n} \sum_{j=1}^{n} \frac{m_i m_j}{r_{ij}^3} \mathbf{R}_i \times \mathbf{r}_{ij} \quad (j \neq i) \tag{1.8}$$

However,

$$\mathbf{R}_i \times \mathbf{r}_{ij} = \mathbf{R}_i \times (\mathbf{R}_j - \mathbf{R}_i) = \mathbf{R}_i \times \mathbf{R}_j$$

and

$$\mathbf{R}_j \times \mathbf{r}_{ji} = \mathbf{R}_j \times (\mathbf{R}_i - \mathbf{R}_j) = -\mathbf{R}_i \times \mathbf{R}_j$$

Therefore, the right-hand side of Eq. (1.8) will sum to zero. Equation (1.8) may then be integrated to yield,

$$\sum_{i=1}^{n} m_i \mathbf{R}_i \times \dot{\mathbf{R}}_i = \mathbf{C}_3, \quad \text{a \textbf{constant} vector} \tag{1.9}$$

The vector $\mathbf{C}_3$ is the normal to the "invariable plane" defined by Laplace, which contains the center of mass.

Equation (1.9) expresses the conservation of the system angular momentum about point O, which is expected since the system of bodies is subject to no net external torque about point O.

Next, because the gravitational force is conservative, a potential energy function V exists such that

$$\mathbf{F}_i = m_i \ddot{\mathbf{R}}_i = -\nabla V = -\frac{\partial V}{\partial \mathbf{R}_i} \tag{1.10}$$

where $-V$ is the work done in assembling the n particles from infinite dispersion to the given configuration. Taking the scalar product of $\dot{\mathbf{R}}_i$ with each side of Eq. (1.10) and summing over all the masses yields

$$\sum_{i=1}^{n} m_i \dot{\mathbf{R}}_i \cdot \ddot{\mathbf{R}}_i = \sum_{i=1}^{n} -\frac{\partial V}{\partial \mathbf{R}_i} \cdot \dot{\mathbf{R}}_i = -\frac{dV}{dt} \tag{1.11}$$

By defining the kinetic energy as

$$T = \tfrac{1}{2} \sum_{i=1}^{n} m_i \dot{\mathbf{R}}_i \cdot \dot{\mathbf{R}}_i \tag{1.12}$$

one recognizes the left-hand side of Eq. (1.11) as dT/dt.

Thus Eq. (1.11) tells us that

$$\frac{dT}{dt} + \frac{dV}{dt} = 0 \tag{1.13}$$

or, after integration, that the total mechanical energy of the system is conserved:

$$T + V = C_4 \tag{1.14}$$

The kinetic energy expression for T is given by Eq. (1.12) and the potential energy that corresponds to the force $\mathbf{F}_i$ in Eq. (1.4) is

$$V = -\frac{G}{2} \sum_{i=1}^{n} \sum_{j=1}^{n} \frac{m_i m_j}{r_{ij}} \qquad (j \neq i) \tag{1.15}$$

We have thus found 10 constants of the motion: The three components of each of the vectors $\mathbf{C}_1$, $\mathbf{C}_2$, and $\mathbf{C}_3$ plus the total mechanical energy C_4. The equations which express the fact that functions of the position and velocity variables are equal to a constant, such as Eq. (1.9). are referred to as "integrals" of the motion. It has been proven that no additional integrals exist for the n-body problem.

The solution of a system of k first-order differential equations,

$$\dot{x}_i = f_i(x_1, x_2, \cdots, x_k, t) \ , \ i = 1, 2, \cdots, k$$

is determined if k distinct (algebraically independent) integrals of the motion

$$g_i(x_1, x_2, \cdots, x_k, t) = \text{constant} \ , \ i = 1, 2, \cdots, k$$

can be found (cf. Whittaker, Ref. 1.2). In such a case the system is said to be *soluble by quadratures*. The equations of motion of the system of n bodies, Eq. (1.4), consist of $3n$ second-order differential equations which are equivalent to $6n$ first-order differential equations (see Prob. 1.21). Thus $6n$ integrals are required to solve for the absolute motion of the n-body system. Since only 10 integrals of the motion exist, even the absolute motion of a system of two bodies cannot be determined in closed form.

1.3 Justification of the Two-Body Model

In the remainder of this book, we will devote a great amount of attention to a special case of the n-body problem, namely the relative motion of two mass particles, the so-called two-body problem. At first glance this appears to be an unrealistic model of the real world for two reasons: (1) in most orbital mechanics problems we encounter there are more than two bodies involved (i.e., a satellite orbiting the earth experiences gravitational attraction from the moon and the sun, as well as other planets), and (2) most celestial bodies hardly seem like mass particles. The reader of this sentence is at this very moment on or near the surface of a very large celestial object whose dimensions are significantly larger than infinitesimal.

The application of a two-body model in a many-body environment is a topic that is treated in more detail in Chap. 7. The basic justification, however, is based on the fact that we can often approximate the motion of our orbiting spacecraft by considering only two bodies: the spacecraft and whatever celestial body has the dominant gravitational effect on the spacecraft. Effects due to other celestial bodies can then be considered to be small perturbations in our two-body model.

The fact that most celestial bodies are not really mass particles (especially when viewed up close!) appears to be a serious shortcoming of the simple two-body model. By analyzing the gravitational force due to a finite sphere of mass, we will be able to alleviate this apparent deficiency in our simple model.

Consider a (hollow) thin spherical shell of radius R as shown in Fig. 1.3. A particle of mass m_2 is located a distance r from the center of the shell as shown. The potential energy due only to the differential mass dm on the shell is obtained from Eq. (1.15) for $n = 2$ and is given by:

$$dV = -\frac{Gm_2\,dm}{\rho} \tag{1.16}$$

The area of the annulus of width $R\,d\theta$ and radius $R\sin\theta$ on the shell defined by all points the same distance ρ from m_2 is given by $2\pi R^2 \sin\theta\,d\theta$. Letting σ denote the mass density per unit area of the shell, the mass of the annulus can be written as $m_a = 2\pi\sigma R^2 \sin\theta\,d\theta$ and, because all the mass is the same distance ρ from m_2, the potential energy due to the annulus is

$$dV = -\frac{2\pi R^2\,G\,\sigma m_2 \sin\theta\,d\theta}{\rho} \tag{1.17}$$

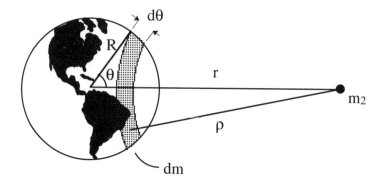

Fig. 1.3. Thin Spherical Shell

The potential energy due to the entire spherical shell is obtained by integrating Eq. (1.17) over the surface of the shell:

$$V_s = -2\pi R^2 \sigma G m_2 \int_0^\pi \frac{\sin\theta}{\rho} d\theta \tag{1.18}$$

To transform the integrand of Eq. (1.18) into a more useful form, one utilizes the fact that $\rho^2 = R^2 + r^2 - 2rR\cos\theta$ which leads to $2\rho\, d\rho = 2rR\sin\theta\, d\theta$ and $\sin\theta\, d\theta/\rho = d\rho/Rr$. Note that $d\rho/d\theta \geq 0$.

Thus at an arbitrary point *outside* the spherical shell ($r \geq R$) the potential energy is

$$V_s = -\frac{2\pi R^2 G \sigma m_2}{Rr} \int_{r-R}^{r+R} d\rho = -\frac{4\pi R^2 G \sigma m_2}{r} \tag{1.19}$$

Noting that the mass of the shell m_s is $4\pi R^2 \sigma$, Eq. (1.19) becomes

$$V_s = -\frac{G m_s m_2}{r} \tag{1.20}$$

Last, one considers a solid sphere of mass m_1 and radius R composed of concentric spherical shells that entirely fill in the original shell of radius R. Because the potential energy due to each shell has the same form (1.20), the potential energy at ($r \geq R$) due to the entire solid sphere is

$$V_o = -\frac{G m_1 m_2}{r} \tag{1.21}$$

where m_1 is simply the sum of the masses of the shells, which is the mass of the solid sphere.

Equation (1.21) has exactly the same form as Eq. (1.15) for $n = 2$! In other words, the surprising result is that the potential energy and hence the gravitational force at any point *outside* a solid sphere is the same as if all the mass were concentrated in a particle at the center of the sphere! Thus the mass particle model is exact for a solid sphere, and we can consider spherical celestial bodies as mass particles, as long as we remember that the distance from the mass particle cannot become less than the radius of the actual spherical body. This important result was first derived by Newton, but his path to the solution was considerably less straightforward than ours — he first had to invent integral calculus!

One final aspect of this problem concerns the mass distribution within the solid sphere. When we discussed forming a solid sphere by stacking concentric shells, nothing was said about the mass densities of the different shells. If the mass density is the same for all the shells, then we have a homogeneous sphere with uniform mass density throughout. The potential energy result obtained above certainly applies in this case. However, it also applies in the more general case that each spherical shell has a different mass density. The potential energy result is still valid because the mass of the solid sphere is still the sum of the masses of the shells. Thus the only requirement on the mass distribution within our solid sphere is that it be spherically symmetric.

1.4 The Two-Body Problem

Fortunately, much may be determined about the motion of the masses in the two-body and certain special cases of few-body problems, especially when it is the *relative motion* that is desired. In this section we will solve for the relative motion in the two-body problem. The solution for the allowed relative motion was first obtained by Isaac Newton in 1683. Newton was arguably the first person capable of obtaining it since the solution requires a law of gravitation, a law of motion, and differential calculus, all of which he invented.

From Eq. (1.4),

$$\ddot{\mathbf{R}}_2 - \ddot{\mathbf{R}}_1 = \ddot{\mathbf{r}}_{12} = \frac{Gm_1}{r_{12}^3}\mathbf{r}_{21} - \frac{Gm_2}{r_{12}^3}\mathbf{r}_{12} \qquad (1.22)$$

or

$$\ddot{\mathbf{r}}_{12} + \frac{G(m_1 + m_2)}{r_{12}^3}\mathbf{r}_{12} = 0 \qquad (1.23)$$

or

$$\ddot{\mathbf{r}} + \frac{\mu}{r^3}\mathbf{r} = 0 \qquad (1.24)$$

where $\mu \equiv G(m_1 + m_2)$ and the subscript notation is dropped because it is no longer necessary.

The definition of μ as the gravitational constant for the two-body system completely characterizes the system. For this reason one does not have to explicitly know the individual masses of the bodies or use the numerical value of G. Also, in many applications the mass of the central body is much larger than the orbiting mass ($m_1 \gg m_2$), and in this case μ is essentially Gm_1. Thus each celestial body has its own value of μ that can be catalogued.

Equation (1.24) for the position $\mathbf{r}$ of m_2 relative to m_1 is nonlinear, but several constants of the motion exist, as discussed in Sect. 1.2, which provide information about the solution without actually obtaining it. For example, taking the vector product

The n-Body Problem

$$\mathbf{r} \times \ddot{\mathbf{r}} + \mathbf{r} \times \frac{\mu}{r^3}\mathbf{r} = 0 \tag{1.25}$$

which may be simply integrated to give

$$\mathbf{r} \times \dot{\mathbf{r}} = \mathbf{h} = \text{a \textbf{constant} vector} \tag{1.26}$$

The vector $\mathbf{r}$ is then normal to the constant vector $\mathbf{h}$. This implies that the relative motion lies in a fixed plane in space called the *orbit plane*, with $\mathbf{h}$ as its normal vector.

Now, to solve Eq. (1.24) take the cross product with the constant vector $\mathbf{h}$:

$$\ddot{\mathbf{r}} \times \mathbf{h} = \frac{-\mu}{r^3}\mathbf{r} \times \mathbf{h} = \frac{-\mu}{r^3}\mathbf{r} \times (\mathbf{r} \times \dot{\mathbf{r}})$$

$$= \frac{\mu}{r^3}[\dot{\mathbf{r}}(\mathbf{r} \cdot \mathbf{r}) - \mathbf{r}(\mathbf{r} \cdot \dot{\mathbf{r}})] \tag{1.27}$$

or

$$\ddot{\mathbf{r}} \times \mathbf{h} = \mu \left[\frac{\dot{\mathbf{r}}}{r} - \frac{\mathbf{r}\dot{r}}{r^2} \right] \tag{1.28}$$

Now note that

$$\frac{d}{dt}\left[\frac{\mathbf{r}}{r}\right] = \frac{r\dot{\mathbf{r}} - \mathbf{r}\dot{r}}{r^2}$$

Therefore, Eq. (1.28) becomes

$$\ddot{\mathbf{r}} \times \mathbf{h} = \mu \frac{d}{dt}\left[\frac{\mathbf{r}}{r}\right] \tag{1.29}$$

which may be integrated directly to yield

$$\dot{\mathbf{r}} \times \mathbf{h} = \mu \left[\frac{\mathbf{r}}{r} + \mathbf{e}\right] \tag{1.30}$$

where $\mathbf{e}$ is a dimensionless vector constant of integration. Because $\mathbf{e}$ is normal to $\mathbf{h}$, $\mathbf{e}$ must lie in the orbit plane. Taking the dot product of $\mathbf{r}$ with Eq. (1.30) yields a scalar equation:

$$\mathbf{r} \cdot (\dot{\mathbf{r}} \times \mathbf{h}) = \mathbf{r} \cdot \mu \left[\frac{\mathbf{r}}{r} + \mathbf{e}\right]$$

or

$$(\mathbf{r} \times \dot{\mathbf{r}}) \cdot \mathbf{h} = h^2 = \mu (r + \mathbf{r} \cdot \mathbf{e}) = \mu (r + re \cos f) \quad (1.31)$$

where the angle f is defined as the angle between $\mathbf{r}$ and $\mathbf{e}$. Solving for r yields:

$$r = \frac{h^2/\mu}{1 + e \cos f} \quad (1.32)$$

which is the equation of a *conic section* in polar coordinates with the origin of the coordinate frame at the *focus* of the conic section. From Eq. (1.32) we see that r will have its minimum value when $f = 0$, that is, the vector $\mathbf{e}$ represents the direction of minimum separation distance.

Equation (1.32) represents a conic section because it is exactly the same equation which results from the formal definition of a conic section: The locus of a point P along which the ratio of a length r (from P to a fixed point F, called the *focus*) to a length w (from P to a fixed line, called the *directrix*) is a constant e, called the *eccentricity*.

Note that we have succeeded in obtaining a closed-form solution to the nonlinear equation of motion (1.24). However, the independent variable in the solution is not time, but the polar angle f, which is called the *true anomaly*. The good news is that we now have a geometrical description of the orbit; one can calculate r for all values of f if the constants μ, h, and e are given. The bad news is that we have lost track of where the orbiting mass is at a specified time. This will eventually come back to haunt us. The missing time information is also evident in the fact that, although the solution to Eq. (1.24) requires six integration constants, our two vector constants $\mathbf{h}$ and $\mathbf{e}$ provide only five *independent* constants due to the fact that $\mathbf{h} \cdot \mathbf{e} = 0$.

It is interesting to examine the effect of the (nonnegative) value of the eccentricity e on the orbit geometry described by Eq. (1.32). Apparently $e = 0$ corresponds to a *circular orbit*, since r is constant in this case. Because an elliptic orbit is bounded, that is, it has a finite maximum radius, it must be that $e < 1$ in order that the denominator of Eq. (1.32) not vanish for $0 \le f < 2\pi$. By considering a circular orbit as a special case of the family of elliptic orbits, the range of values of eccentricity for elliptic orbits is $0 \le e < 1$.

If $e = 1$ in Eq. (1.32) the value of $r \to \infty$ as $f \to \pm \pi$, which describes a *parabolic orbit*, and if $e > 1$, the value of $r \to \infty$ along asymptotes defined by values of $f = f_\infty < \pi$ given by $e \cos f_\infty = -1$, which describes a *hyperbolic orbit*. These geometrical aspects will be investigated further in the next sections. These conic orbits composed of elliptic, parabolic, and hyperbolic orbits describe the possible motions in our two-body problem.

The n-Body Problem

It may seem as if we have arrived at a contradiction, in that in Sec. 1.2 we observed that even the two-body problem could not be solved completely due the limited number of constants of the motion. However, the solution we have obtained, as Eq. (1.32), describes only the path of the *relative* motion. As seen by an inertial observer, the motion would be more complex. While each body moves in an elliptic (for example) path as seen from the other, the system center of mass translates with constant velocity and hence the bodies will simultaneously orbit the system center of mass. Fortunately it is the relative motion that most interests us, for example, in the motion of a planet relative to the sun. Exactly what the motion of our sun is relative to other parts of our galaxy or other parts of the universe is of little importance for missions within our solar system.

1.5 The Elliptic Orbit

The fact that an elliptic orbit is one solution for the relative motion of two bodies verifies Kepler's First Law. It will be more useful to consider the elliptic orbit first because (1) it is comparatively simple to visualize, (2) this is the type of orbit of most natural and artificial satellites in the solar system, and (3) some interesting things are learned along the way. We have found the separation in the relative motion to be governed by Eq. (1.32). The polar equation of an ellipse is

$$r = \frac{a(1-e^2)}{1+e \cos f} \quad (1.33)$$

where a represents the *semimajor axis*. Both it, the *semiminor axis* b, and the true anomaly f are shown in Fig. 1.4.

Comparing Eqs. (1.32) and (1.33) yields the *specific angular momentum*, h, as a function of the masses and orbit geometry:

$$h = [\mu a (1-e^2)]^{\frac{1}{2}} \quad (1.34)$$

Three points in the orbit are of special interest. When $f = 0$, eqn. (1.33) yields $r = a(1-e)$. This is the minimum separation for the satellite and attracting body and the satellite is said to be at *periapse*. For a satellite of the earth, this point is referred to as the *perigee*, for a satellite of the moon it is called *perilune*, and so on. When $f = \pi/2$, Eq. (1.33) yields $r = a(1-e^2)$. This distance is referred to as the *semi-latus rectum* or simply the *parameter* of the ellipse and is usually designated by p. Equation (1.33) may then also be written as

$$r = \frac{p}{1+e \cos f} \quad (1.35)$$

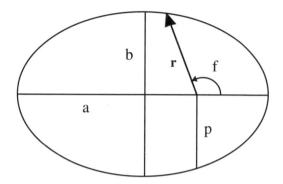

Fig. 1.4. Geometry of the Elliptic Orbit

and from Eq. (1.35) the specific angular momentum h and the parameter p are related by

$$p = \frac{h^2}{\mu} \tag{1.36}$$

When $f = \pi$ Eq. (1.33) gives $r = a(1+e)$. This is the greatest separation possible for the two bodies. In this case the satellite is said to be at *apoapse* (or apogee, apolune, etc.).

With reference to Fig. 1.5, which portrays the motion of m_2 as seen by an observer on m_1, we see that

$$\dot{\mathbf{r}} = \dot{r}\hat{i} + r\dot{f}\hat{j} = \mathbf{v} \tag{1.37}$$

where the unit vectors rotate with the radius vector. Then, from Eq. (1.32) we have

$$\mathbf{h} = \mathbf{r} \times \dot{\mathbf{r}} = r\hat{i} \times (\dot{r}\hat{i} + r\dot{f}\hat{j}) = r^2 \dot{f} \hat{k} \tag{1.38}$$

or $h = |\mathbf{h}| = r^2 \dot{f} = $ constant. Note also, however, that the differential element of area swept out by the radius vector as it rotates through an angle df is, to first order, $dA = \frac{1}{2} r^2 df$. Therefore, this implies that

$$\frac{dA}{dt} = \frac{1}{2} r^2 \frac{df}{dt} = \frac{h}{2} = \text{constant} \tag{1.39}$$

That is, the rate at which the radius vector sweeps out area (the *areal velocity*) is a constant, and this follows directly from the fact that the orbital angular momentum is conserved. This verifies Kepler's Second Law. In

The n-Body Problem   17

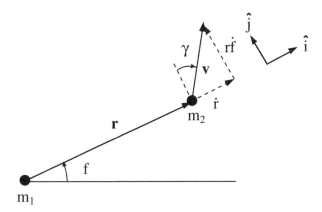

Fig. 1.5. Unit Vector Definitions

addition, and as a consequence, the time required for one complete orbit, the *orbital period T* is

$$T = \frac{\text{enclosed area of the ellipse}}{dA/dt} \qquad (1.40)$$

or

$$T = \frac{\pi ab}{\dot{A}} = \frac{2\pi ab}{h} = \frac{2\pi a \cdot a(1-e^2)^{1/2}}{(\mu a)^{1/2}(1-e^2)^{1/2}} = 2\pi\left[\frac{a^3}{\mu}\right]^{1/2} \qquad (1.41)$$

where b is the semiminor axis of the ellipse. The curious fact appears that the elliptic orbit period depends only on the semimajor axis and is independent of the eccentricity. Thus the period depends only on the size of the elliptic orbit and is independent of its shape. Equation (1.41) also verifies Kepler's Third Law *if* the value of μ is the same for all planets in the solar system. In this application $\mu = G(m_s + m_p)$ where m_s is the mass of the sun and m_p is a planet mass. Because $m_s \gg m_p$ for all the planets, μ is essentially Gm_s and Kepler's Third Law is correct for the "small" orbiting masses of the planets.

1.6 Parabolic, Hyperbolic, and Rectilinear Orbits

It was found in Sec. 1.4 that the relative motion in the two-body problem was described by

$$r = \frac{h^2/\mu}{1 + e \cos f} \quad (1.32)$$

regardless of the type of conic section represented by the orbit, with the numerator being a constant. For $e < 1$, $e = 1$, $e > 1$ the orbit will be an ellipse, a parabola, or hyperbola, respectively. In addition there are other orbits for which $e = 1$, namely *rectilinear* ellipses, parabolas, and hyperbolas, which are characterized by $p = h = 0$. They are one-dimensional orbits in the form of a straight line on the major axis. The angular momentum h is 0 because the position and velocity vectors are parallel (or antiparallel). The periapse radius for all the rectilinear conic orbits is 0. Although these orbits seem somewhat pathological, they can exist in practice and they can also be used to simplify derivations and to geometrically interpret results as in Sec. 4.4.

Note that the true anomaly f is not useful for rectilinear orbits, since from Eq. (1.39), which is a statement of Kepler's Second Law, $\dot{f} = 0$ in the rectilinear case. This makes sense because there is no area swept out by the radius vector in a one-dimensional orbit. One must use a different variable introduced in Sec. 2.2 to keep track of position along a rectilinear orbit.

For the general elliptic orbit we showed that the constant h^2/μ could be identified as a distance, the parameter p, of the ellipse. For the other conics this remains true, so that the equation

$$r = \frac{p}{1 + e \cos f} \quad (1.35)$$

will represent the parabola and hyperbola equally well. Figure 1.6 depicts the parabolic and hyperbolic orbits.

In the parabolic orbit, only the parameter p characterizes the different parabolic orbits because the eccentricity for all parabolas is unity and the semimajor axis is infinite, as may be seen by considering the parabolic orbit as the limiting case of an elliptic orbit for which the apoapse tends to infinity. The parameter for a parabolic orbit is equal to twice the periapse radius.

For the hyperbolic orbit it is convenient to leave the form of the formulas unchanged and consider the semimajor axis to be negative ($a < 0$). The periapse radius for the hyperbolic orbit is then

$$r_p = (-a)(e - 1) = a(1 - e) \quad (1.42)$$

and the parameter remains

$$p = a(1 - e^2) \quad (1.43)$$

Also shown in Fig. 1.6 (b) are the *turn angle* δ and the *aiming radius* Δ for a hyperbolic orbit.

The n-Body Problem

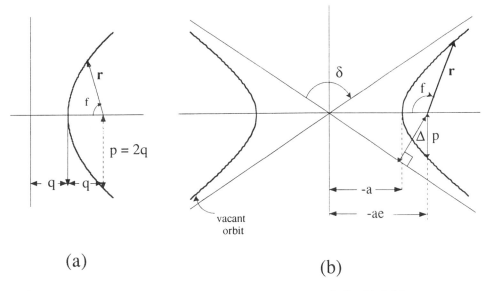

Fig. 1.6. Geometry of Parabolic (a) and Hyperbolic (b) Orbits

1.7 Energy of the Orbit

Determining the energy of each of the orbits provides a simple method for finding the speed of the satellite at any point on the orbit. The total mechanical energy of the satellite is the sum of the kinetic and potential energies,

$$T + V = \frac{mv^2}{2} - \frac{m\mu}{r} \tag{1.44}$$

and is constant because the only force acting on the satellite, the mutual gravitational attraction, is a conservative force. It is convenient to consider the specific energy, which is the energy per unit mass of satellite:

$$\varepsilon = \frac{v^2}{2} - \frac{\mu}{r} = \text{constant}, \tag{1.45}$$

but what is the value of this constant *specific orbit energy*?

At periapse, regardless of the type of conic orbit, the velocity and radius vectors are orthogonal so that, from Eq. (1.38):

$$h = r_p v_p = (\mu p)^{½} \tag{1.46}$$

Employing Eq. (1.42) for r_p yields

$$v_p = \frac{(\mu p)^{1/2}}{r_p} = (\mu p)^{1/2} \frac{(1+e)}{p} = (1+e)\left[\frac{\mu}{p}\right]^{1/2} \qquad (1.47)$$

Therefore, the constant ε in Eq. (1.45) must be equal to

$$\varepsilon = \frac{v_p^2}{2} - \frac{\mu}{r_p} = \frac{(1+e)^2 \mu}{2p} - \frac{(1+e)\mu}{p} \qquad (1.48)$$

We immediately see from this that the energy is given by $\varepsilon = 0$ for a parabolic orbit ($e = 1$, $p \neq 0$).

For elliptic and hyperbolic orbits we have from Eq. (1.36) that $p = a(1 - e^2)$. Substituting this value into Eq. (1.48) yields

$$\varepsilon = \frac{(1+e)^2 \mu}{2a(1-e^2)} - \frac{(1+e)\mu}{a(1-e^2)} = \frac{-\mu}{2a} \qquad (1.49)$$

Thus the total energy is *independent of eccentricity* and depends only on the semimajor axis for all conic orbits. By combining Eqs. (1.45) and (1.48) we can write the velocity at a given radius as

$$v^2 = \mu \left(\frac{2}{r} - \frac{1}{a}\right) \qquad (1.50)$$

Equation (1.50) is known as the *vis–viva equation* and is very useful in orbit mechanics calculations.

For an elliptic orbit ($a > 0$), the specific energy of Eq. (1.49) is negative, implying that the satellite is bound to the attracting center since, from Eq. (1.50), for $r > 2a$, v^2 would have to be negative, which is not possible because v is real-valued. For a hyperbolic orbit ($a < 0$) the specific energy is positive and v^2 remains positive as $r \to \infty$ in Eq. (1.50). The satellite will thus proceed indefinitely far away from the attracting center.

For a parabolic orbit ($1/a = 0$) we see that Eq. (1.50) is also valid with $v^2 \to 0$ as $r \to \infty$, that is the satellite has just enough total energy to escape from the attracting center to infinite radius. For this reason the velocity required at a given point in space to be on a parabolic trajectory is called *escape velocity*, and is denoted by v_e.

Example 1.1

Find expressions for the escape velocity v_e and circular orbit velocity v_c at a point a distance r from the earth's center.

To be in a parabolic orbit requires that

$$\varepsilon = 0 = \frac{v^2}{2} - \frac{\mu}{r}$$

and therefore, $v_e = (2\mu/r)^{1/2}$ is the escape velocity.
To be in a circular orbit of radius r requires that $a = r$:

$$\varepsilon = \frac{-\mu}{2a} = \frac{-\mu}{2r} = \frac{v^2}{2} - \frac{\mu}{r}$$

and therefore, $v_c = (\mu/r)^{1/2}$ is the circular orbit velocity. To escape thus requires $\sqrt{2}$ times the circular orbit speed at a given radius.

In the case of a hyperbolic orbit the excess kinetic energy over what is needed to escape the center of attraction is described in terms of the *hyperbolic excess speed* v_∞, defined as the speed at infinite radius. From Eq. (1.50):

$$v_\infty = \left[\frac{\mu}{-a}\right]^{1/2} \qquad (1.51)$$

which is real-valued because $a < 0$ for a hyperbolic orbit. This hyperbolic excess speed provides a much more meaningful description of the total energy of a hyperbolic orbit than the value of its semimajor axis.

Table 1.1 Name that Conic

	$0 \le e < 1$	$e = 1$	$e > 1$
$1/a < 0$		rectilinear hyperbola	hyperbola
$1/a = 0$		parabola (including rectilinear)	
$1/a > 0$	ellipse	rectilinear ellipse	

The key results of Secs. 1.6 and 1.7 are summarized in Table 1.1. Note that the eccentricity increases as one moves left to right in Table 1.1, and the *total energy* of Eq. (1.49) increases as one moves vertically upward through the table.

References

1.1 Prussing, J. E., "Kepler's Laws of Planetary Motion," *The Encyclopedia of Physics* 3rd Edition, R. Besancon, ed. Van Nostrand Reinhold Co., New York, 1985 (reprinted with permission)

1.2 Whittaker, E. T., *A Treatise on the Analytical Dynamics of Particles and Rigid Bodies*, 4th Edition, Cambridge University Press, London, England, 1965.

Problems

1.1 Consider the relative motion of two equal mass particles.
a) Determine the required constant angular velocity ω_0 of the relative position vector in order to maintain a constant separation distance r_0.
b) Determine the value of ω_0 for unequal masses m_1 and m_2.

1.2 Consider three equal mass particles in an equilateral triangle configuration. Determine the required constant angular velocity ω_0 to maintain a constant separation distance d_0 between the masses.

1.3 (R. H. Battin)
a) Derive the equation of motion of a mass particle m_2 with respect to the barycenter (center of mass) of a two-body system. The solution should be expressed in terms of only the position vector relative to the barycenter, the universal gravitational constant, and the masses m_1 and m_2.
b) Specialize your result to the case $m_1 \gg m_2$.

1.4* (A. E. Roy)
a) Show that if a solution to the n-body problem of Eq. (1.4) were known, other solutions could be generated by scaling all lengths by a factor D and time by the factor $D^{3/2}$
b) Is this consistent with Kepler's Third Law for the case $n = 2$?

1.5 a) Show that on a parabolic orbit the flight path angle γ (See Fig. 1.5) is equal to half the true anomaly.
b) Show that this property explains the operation of a parabolic antenna or mirror, namely, that all incoming rays parallel to the major axis are reflected to the focus of the parabola.

The n-Body Problem

1.6 By modifying the derivation of the gravitational potential at a point *outside* a spherical shell of uniform mass density, obtain the potential *and* the associated gravitational force at an arbitrary point *inside* a spherical shell.

1.7 a) Determine the gravitational force at an arbitrary point *inside* a solid homogeneous sphere.
b) Obtain the differential equation of motion of a mass particle that is free to slide frictionlessly in a straight-line tunnel through the center of the sphere.
c) Solve the equation of motion and determine the period of oscillation of the particle.
d) Calculate the radius of the circular orbit about the sphere that has the same period as the particle in the tunnel. Express your answer in terms of the radius of the sphere.

1.8* Obtain the equation of motion of a mass particle in a frictionless straight-line tunnel connecting two *arbitrary* points on the surface of a solid homogeneous sphere.

1.9 a) In the derivation of Eq. (1.29) the fact is used that $r\dot{r} = \mathbf{r} \cdot \dot{\mathbf{r}}$. Does the variable $\dot{r}$ represent dr/dt or the magnitude of $\dot{\mathbf{r}}$?
b) Calculate the radial component of $\dfrac{d}{dt}(\dfrac{\mathbf{r}}{r})$.

1.10 At time t_0 in units for which $\mu = 1$ (so-called canonical units), the following data is given for a two-body problem:

$$\mathbf{r}^T = [\,0\ 2\ 0\,] \text{ and } \mathbf{v}^T = [\,-1\ 1\ 0\,]/\sqrt{2}$$

a) Calculate the vectors $\mathbf{h}$ and $\mathbf{e}$ and verify that $\mathbf{h} \cdot \mathbf{e} = 0$.
b) Write the polar equation $r(f)$ for this conic orbit.
c) What is the value of f at time t_0?
d) Determine the speed of the spacecraft when $r = 32$.
e) Determine the value of f when $r = 32$.
f) For the earth $\mu = 3.986 \times 10^5$ km^3 s^{-2}. Determine the value in kilometers of the length unit (LU) for which $\mu = 1$ LU^3 s^{-2}.

1.11 a) Derive an expression for the eccentricity e in terms of the initial speed v_0, radius r_0, and flight path angle γ_0.
b) Discuss the special cases: (1) $r_0 v_0^2 / \mu = 2$, and (2) $r_0 v_0^2 / \mu = 1$.
c) Obtain a general expression for the parameter p.

1.12 a) Determine the location of the point(s) on an elliptic orbit at which the speed is equal to circular orbit speed.
b) Obtain an expression for the flight path angle at this location.

1.13 Tracking data for an earth satellite indicates altitude = 600 km, $r\dot{f} = 7$ km/s, and $\dot{r} = 3.5$ km/s. Determine the eccentricity e and the true anomaly f at the data point (earth radius = 6378 km).

1.14 In terms of v_s, which is circular orbit speed at the surface of a planet of radius r_s, and r_p, which is periapse radius of a hyperbolic orbit about the planet, show that

a) $\mu = r_s v_s^2$ b) $e = 1 + \psi$; $\psi \equiv \left(\dfrac{v_\infty}{v_s}\right)^2 \left[\dfrac{r_p}{r_s}\right]$

c) $\cdot\sin(\dfrac{\delta}{2}) = \dfrac{1}{1+\psi}$

d) $\dfrac{\Delta}{r_s} = \dfrac{r_p}{r_s}\sqrt{1 + 2/\psi}$

1.15* For an elliptic orbit
 a) Determine the location in the orbit where the flight path angle is a maximum.
 b) Determine the value of the maximum flight path angle.
 c) Obtain the value of the orbital speed at the point of maximum flight path angle.
 Hint: Use Lagrange multipliers (see Sec. 5.6) to maximize γ subject to the constraints: $rv\cos\gamma - h = 0$ and $v^2/2 - \mu/r + \mu/2a = 0$.

1.16 In a different universe, a pair of particles move through space under the action of a law of gravitation for which the force of attraction between the particles is directly proportional to the product of their masses and *directly proportional* to their separation distance. Show that the orbit of one body about the other will be an ellipse with the other body at the *center* of the ellipse.

1.17 If m represents the mass of a planet and M represents the mass of the sun, then the period T of the planet orbit about the sun is given by

$$T^2 = \dfrac{4\pi^2 a^3}{G(M+m)}$$

A similar expression describes the period of the orbit of a satellite about the planet. Use these relationships to show that if the periods and semimajor axes of the orbits of the planet and the planetary satellite are known, then the mass of the planet (in units of the sun's mass) can be found.

1.18 a) Express the vector equation of relative motion of two bodies, Eq. (1.24), in component form using polar coordinates.
b) Determine the linear, constant-coefficient differential equation of motion that results if one utilizes true anomaly, rather than time, as the independent variable along with the reciprocal of the magnitude of the radius as the dependent variable ($u = 1/r$).

1.19* (R. H. Battin)
Consider two particles of masses m_1 and m_2 in orbit about their common center of mass. Show that if each is moving in an elliptic orbit, the semimajor axes of the two orbits are in inverse ratio to their masses and their eccentricities are the same.

1.20* Show that the potential energy function V of Eq. (1.15) yields the force vector $\mathbf{F}_i$ in Eq. (1.4) by proving that $\mathbf{F}_i = -\partial V / \partial \mathbf{R}_i$.

1.21 a) Show that the equation of motion for the harmonic oscillator, $\ddot{y} + y = 0$, may be written as a system of two first-order differential equations,

$$\dot{x}_1 = x_2$$
$$\dot{x}_2 = -x_1$$

b) Show that this system has two integrals of the motion,

$$x_1^2 + x_2^2 = c_1$$
$$\tan^{-1}\frac{x_1}{x_2} - t = c_2$$

and solve these equations for $x_1(t)$ and $x_2(t)$.

2

Position in Orbit as a Function of Time

2.1 Introduction

The determination of the position of a spacecraft along its orbit at a given time is clearly a practical problem. It might seem to be a trivial one, inasmuch as we have already derived expressions for the magnitude of the radius and for the (constant) areal velocity. However, the invariability of the latter implies that for an eccentric orbit the angular velocity of the spacecraft is continuously varying, and unfortunately the expression for the position angle f (as the integral of $\dot{f}$), which yields angular position versus time, is extremely cumbersome and impractical to use. We can approach the problem in a more direct way, in the manner originally used by Kepler.

2.2 Position and Time in an Elliptic Orbit

Figure 2.1 shows an ellipse with semimajor axis a inscribed within a circle of radius a. With point O the center of both figures and the origin of the coordinate system shown, the equations describing the conic sections are:

$$\frac{x^2}{a^2} + \frac{y^2}{b^2} = 1$$

for the ellipse, and

$$\frac{x^2}{a^2} + \frac{y^2}{a^2} = 1$$

for the circle.

Thus for a given x coordinate, the corresponding y coordinates are:

$$y = \frac{b}{a}\left[a^2 - x^2\right]^{1/2}$$

on the ellipse, and

$$y = \left[a^2 - x^2\right]^{1/2}$$

on the circle, so that the magnitude of the y coordinate of any point on the ellipse is smaller than that of the corresponding point on the circle by the factor b/a.

The time t, required to travel from the close approach point, or *periapse*, through the true anomaly f along the orbit is given by Kepler's Second Law as:

$$\frac{t}{T} = \frac{\text{area of sector FVP}}{\text{area of ellipse}} \tag{2.1}$$

where T represents the orbit period. However,

$$\text{area FVP} = \text{area PSV} - \text{area PSF}$$

$$= \text{area PSV} - \frac{1}{2}(ae - a\cos E)(a\sin E)\frac{b}{a} \tag{2.2}$$

where E is an auxiliary angle, called the *eccentric anomaly*, defined in Fig. 2.1. But,

$$\text{area PSV} = \frac{b}{a}\text{area QSV} = \frac{b}{a}\left[\tfrac{1}{2}a^2 E - \tfrac{1}{2}a\cos E\, a\sin E\right] \tag{2.3}$$

Therefore,

$$\text{area FVP} = \tfrac{1}{2}ba(E - e\sin E) \tag{2.4}$$

so that, employing Eq (1.41),

$$t = \frac{\tfrac{1}{2}ba(E - e\sin E)}{\pi ab}\, 2\pi\left(\frac{a^3}{\mu}\right)^{1/2} = \left(\frac{a^3}{\mu}\right)^{1/2}(E - e\sin E) \tag{2.5}$$

The mean angular rate of the satellite, symbolized by n, and called the *mean motion*, is simply

$$n = \frac{2\pi}{T} = \left[\frac{\mu}{a^3}\right]^{1/2} \tag{2.6}$$

Therefore, we may define an auxiliary angle $M = nt$, called the *mean anomaly*, which represents physically the angular displacement of a fictitious satellite that travels at the mean angular rate n as opposed to the true rate $\dot{f}$.

In terms of mean anomaly Eq. (2.5) can be written as

$$M = E - e\sin E \tag{2.7}$$

Equation (2.7) is known as *Kepler's equation* and relates time (in terms of mean anomaly) to position (in terms of eccentric anomaly) for an elliptic orbit. In our definition of mean anomaly we have assumed that $t = 0$ at periapse. In the case of an arbitrary time reference, t is replaced by the elapsed time since periapse passage $t - \tau$ where τ is the *time of periapse*

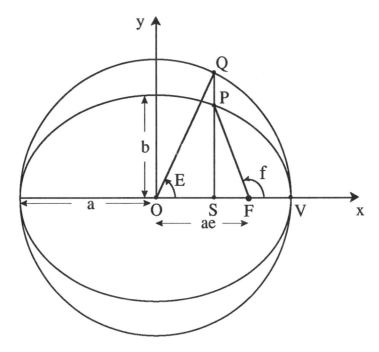

Fig. 2.1 Definition of Eccentric Anomaly

passage. The value of τ represents the *sixth* independent constant, in addition to the vectors **h** and **e**, needed for a unique solution to the two-body problem, as discussed in Sec. 1.4. As a consequence of Kepler's second law, the mean anomaly will lag behind the true anomaly in the first and second quadrants. The situation will then reverse in the third and fourth quadrants. As shown in Prob. 2.4 the mean and eccentric anomalies exhibit the same property.

Unfortunately, Eq. (2.7) still does not provide position, f, as a function of t. We need to relate true anomaly f to eccentric anomaly E. With reference again to Fig. 2.1, we see that the length of line segment OS is

$$OS = a \cos E = ae + r \cos f$$

or, employing Eq. (1.33) for r yields

$$a \cos E = ae + \frac{a(1-e^2)\cos f}{1+e \cos f} \qquad (2.8)$$

or

$$\cos E = \frac{e + \cos f}{1 + e \cos f} \qquad (2.9)$$

Utilizing Eq. (2.9)

$$\tan^2 \frac{E}{2} = \frac{1-\cos E}{1+\cos E} = \left[\frac{1-e}{1+e}\right]\left[\frac{1-\cos f}{1+\cos f}\right] \qquad (2.10)$$

which yields

$$\tan \frac{E}{2} = \left[\frac{1-e}{1+e}\right]^{1/2} \tan \frac{f}{2} \qquad (2.11)$$

Equation (2.11) is a very convenient conversion formula that alleviates quadrant ambiguity (see Prob. 2.1). It also follows directly from Eq. (2.9) that

$$r = \frac{a(1-e^2)}{1+e\cos f} = a(1-e\cos E) \qquad (2.12)$$

which describes the radius directly in terms of eccentric anomaly.

Kepler's Eq. (2.7) has many applications in orbital mechanics that fall into two cases: (1) For a given elliptic orbit, determine the time at which the orbiting body will be at a specified position in the orbit, and (2) determine the position of the orbiting body at a specified time. These two cases sound similar, but are quite different both in terms of the physical problems they represent and the degree of difficulty in solving Kepler's equation.

In Case 1, the solution to Eq. (2.7) is straightforward. The value of E corresponding to the specified position in the orbit is calculated, such as by using Eq. (2.11). The time is then calculated directly from Eq. (2.7). An application of this case is the determination of the time at which an earth satellite passes from sunlight into the earth's shadow. The location of this point is known from the geometry. The time at which the satellite passes into and out of the earth's shadow determines the amount of time during each orbit electrical power must be drawn from batteries and the amount of time solar cells can recharge the batteries.

In Case 2 on the other hand, the solution of Kepler's equation is much more difficult because the equation is transcendental in the unknown variable E. One cannot write a closed-form expression for E as a function of the time t, although many famous mathematicians have tried over the years. However, there are important situations in which the location of a satellite at a particular time must be known. Examples include sending and receiving radio signals and performing a rendezvous with an orbiting space station. There are various approximations and series expansion methods that can be used, but numerical iteration, as discussed in the next section, is commonly employed to obtain a solution.

When one is faced with solving a transcendental equation, the questions of existence and uniqueness of a solution must be addressed. It would be silly to attempt to iteratively determine a real-valued solution to an equation that does not have one, for example $x^2 + \sin x + 8 = 0$. Also, if a solution does exist, one would like to know if it is unique because an iterative technique might converge to a solution other than the one we are seeking. A simple example of an equation having a nonunique solution is $e^{-x} - \cos x = 0$.

Fortunately, it is relatively easy to show that a solution to Kepler's equation for eccentric anomaly E exists and is unique. To do this, rewrite Eq. (2.7) in the form

$$F[E] = E - e \sin E - M = 0 \qquad (2.13)$$

The solution for E is then a zero of the nonlinear function $F[E]$. Suppose that the specified value of time t corresponds to a value of mean anomaly M that is in the interval

$$k\pi \leq M < (k+1)\pi \qquad (2.14)$$

for some integer k. In terms of the function F in Eq. (2.13) $F[k\pi] = k\pi - M \leq 0$ and $F[(k+1)\pi] = (k+1)\pi - M > 0$. Because the function F is continuous, this implies that the function vanishes at least once in the interval $k\pi \leq E < (k+1)\pi$. The fact that the slope of the function given by

$$F'[E] = \frac{dF}{dE} = 1 - e \cos E$$

is always positive for $0 \leq e < 1$ guarantees that there is only one zero of $F[E]$ in the interval. It is interesting to note that the slope $F'[E]$ is equal to r/a [see Eq. (2.12)] and thus varies between a minimum value of $1 - e$ at periapse and a maximum value of $1 + e$ at apoapse. This fact is useful in obtaining bounds on the solution.

2.3 Solution for the Eccentric Anomaly

As stated earlier, Eq. (2.7), which needs to be solved for E, is transcendental and has no closed-form solution in terms of elementary functions. It is customarily solved numerically by some method of successive approximations, the most familiar of which is Newton's method. However, many such methods exist and there is a substantial literature on the application of these methods to the Kepler problem, as well as to the choice of a first guess or starting value that all such iteration methods require.

We will use the familiar Newton method in the following example. To find a root of the equation $F(x) = 0$, one generates, for a current approximation x_k, a recursion formula for the next approximation x_{k+1} using a straight-line approximation as

Position in Orbit as a Function of Time 31

$$x_{k+1} = x_k - \frac{F(x_k)}{F'(x_k)} \quad (2.15)$$

The iteration is continued until $|F(x_k)|$ is sufficiently close to zero, or until the value $|x_{k+1} - x_k|$ is sufficiently small. Newton's method works well for well-behaved functions (i.e., continuous with continuous, nonzero first derivatives) which, in fact, Kepler's equation (2.13) happens to be.

Example 2.1

An earth satellite orbit has a semimajor axis $a = 4 R_e$ and a perigee (periapse) radius $r_p = 1.5 R_e$, where R_e is earth radius. Find the true anomaly at t = 4 hours after perigee passage.

From the data given the value of $a = 4 R_e = 4$ (6378 km) = 25,512 km and, since $r_p = a(1-e) = 4 R_e (1-e) = 1.5 R_e \to e = 5/8$.

Therefore using $\mu = 3.986 \times 10^5 \text{km}^3 \text{ s}^{-2}$,

$$M = nt = (\mu/a^3)^{1/2} t$$

$$= (1.549 \times 10^{-4} \text{ s}^{-1})(4\text{h} \times 3600 \text{ s/hr})$$

$$= 2.231 \text{ radians} = E - e \sin E$$

which leads to Kepler's Eq. of the form

$$F(E) = E - e \sin E - M = E - 5/8 \sin E - 2.231,$$

along with $F'(E) = 1 - 5/8 \cos E$.

Using Newton's method,

$$E_{i+1} = E_i - \frac{(E_i - 5/8 \sin E_i - 2.231)}{1 - 5/8 \cos E_i}$$

As a first approximation, choose $E_o = M = 2.231$, then

$$E_1 = 2.231 - (-.357) = 2.588$$

$$E_2 = 2.588 - (.0183) = 2.570$$

$$E_3 = 2.570 - (< 10^{-3}) = 2.570$$

and, from Eq. (2.11):

$$\tan \frac{f}{2} = \left[\frac{1+e}{1-e}\right]^{1/2} \tan \frac{E}{2} \to f = 2.861 \text{ radians} = 164°$$

Also, $r = aF' = 38,922$ km.

When the eccentricity is small, the "traditional" starting value $E_o = M$ (which would be the solution if the orbit were circular) is a reasonable first approximation. From Ex. 2.1 it can be seen that even for a significant eccentricity ($e = 5/8$) this starting value can lead to rapid convergence. It can be shown that the solution to Kepler's equation lies within the bounds $M \leq E \leq M + e$ [see Prob. 2.4] and that it may be nearer the upper bound than the lower for much of the orbit. When the eccentricity is large, the use of the lower bound M as the starting value is not a good choice and can delay or prevent the convergence of the iteration [2.1]. An improvement in convergence speed is obtained with a starting value located between the bounds, for example, midway between the bounds at $E_o = M + e/2$. A more refined starting value is given by

$$E_o = \frac{M(1 - \sin u) + u \sin M}{1 + \sin M - \sin u} \tag{2.16}$$

where $u \equiv M + e$. This starting value is a secant interpolation between the values at the bounds $F(M)$ and $F(M + e)$. Another iteration method for a successive approximations solution of Kepler's equation will be described in Sec. 2.5.

2.4 The f and g Functions and Series

Since a knowledge of the position and velocity vectors at any time allows the determination of the constants of integration of the two-body problem (**h** and **e**), as shown in Sec. 1.4, it is clear that **r** and **v** at an arbitrary time determine the orbit. One might suppose then that **r** and **v** at an arbitrary time t can be written as a linear combination of **r** and **v** at an epoch time, t_o. In fact this is easily shown. Assume that

$$\mathbf{r}(t) = f(t, t_o, \mathbf{r}_o, \mathbf{v}_o)\mathbf{r}_o + g(t, t_o, \mathbf{r}_o, \mathbf{v}_o)\mathbf{v}_o \tag{2.17}$$

where $\mathbf{r}_o = \mathbf{r}(t_o)$ and $\mathbf{v}_o = \mathbf{v}(t_o)$. Then,

$$\mathbf{v}(t) = \dot{\mathbf{r}}(t) = \dot{f}(t, t_o, \mathbf{r}_o, \mathbf{v}_o)\mathbf{r}_o + \dot{g}(t, t_o, \mathbf{r}_o, \mathbf{v}_o)\mathbf{v}_o \tag{2.18}$$

With regard to Fig. 2.1, **r** may be written as

$$\mathbf{r} = a(\cos E - e)\hat{i} + \frac{b}{a} a \sin E \hat{j} \tag{2.19}$$

where $\hat{i}$ and $\hat{j}$ are unit vectors in the directions of the positive x and y axes, respectively. Therefore,

$$\mathbf{r} = a[(\cos E - e)\hat{i} + \sqrt{1 - e^2} \sin E \hat{j}] \tag{2.20}$$

Position in Orbit as a Function of Time

and hence $r = |\mathbf{r}| = a(1 - e \cos E)$ (Eq. 2.12.) Taking the first time derivative of Eq. (2.7) yields

$$n = \dot{E}(1 - e \cos E) \tag{2.21}$$

so that

$$\dot{E} = \frac{\sqrt{\mu/a^3}}{(1 - e \cos E)} = \frac{\sqrt{\mu/a}}{r}. \tag{2.22}$$

Taking the time derivative of (2.20) and employing (2.22) yields

$$\mathbf{v} = \frac{d\mathbf{r}}{dt} = \frac{\sqrt{\mu a}}{r}[-\sin E \,\hat{i} + \sqrt{1 - e^2} \cos E \,\hat{j}]. \tag{2.23}$$

Therefore

$$\mathbf{r}_o = a[(\cos E_o - e)\hat{i} + \sqrt{1 - e^2} \sin E_o \,\hat{j}]$$

$$\mathbf{v}_o = \frac{\sqrt{\mu a}}{r_o}[-\sin E_o \,\hat{i} + \sqrt{1 - e^2} \cos E_o \,\hat{j}]. \tag{2.24}$$

Equations (2.24) may be inverted to solve for $\hat{i}$ and $\hat{j}$ with the result,

$$\hat{i} = \cos E_o \frac{\mathbf{r}_o}{r_o} - \sqrt{a/\mu} \sin E_o \,\mathbf{v}_o$$

$$\hat{j} = \frac{\sin E_o}{\sqrt{1 - e^2}} \frac{\mathbf{r}_o}{r_o} + \frac{\sqrt{a/\mu}(\cos E_o - e)}{\sqrt{1 - e^2}} \mathbf{v}_o. \tag{2.25}$$

The unit vectors $\hat{i}$ and $\hat{j}$ from (2.25) may now be substituted back into (2.20) and (2.23) to yield expressions for $\mathbf{r}$ and $\mathbf{v}$ in the form of Eqs. (2.17) and (2.18), respectively. Collecting terms multiplying $\mathbf{r}_o$ and $\mathbf{v}_o$ yields

$$f = \frac{a}{r_o}[(\cos E - e)\cos E_o + \sin E \sin E_o]$$

$$g = \left[\frac{a^3}{\mu}\right]^{1/2} [\sin(E - E_o) - e(\sin E - \sin E_o)]$$

$$\dot{f} = -\frac{\sqrt{\mu a}}{r r_o} \sin(E - E_o)$$

$$\dot{g} = \frac{a}{r}[\cos(E_o - E) - e \cos E]$$

$$= 1 - \frac{a}{r}[1 - \cos(E - E_o)] \tag{2.26}$$

In order to use these "f and g functions" we need the values of the semimajor axis and eccentricity. These may be determined from the initial conditions ($\mathbf{r}_o$, $\mathbf{v}_o$) in a manner that will be described in Sec. 3.3. It is possible to express f and g more simply; since $r_o = a(1 - e\cos E_o)$, the quantity $(-e\cos E_o) = r_o/a - 1$, so that f becomes

$$f = \frac{a}{r_o}\left[\frac{r_o}{a} - 1 + \cos E \cos E_o + \sin E \sin E_o\right]$$

$$= 1 - \frac{a}{r_o}[1 - \cos(E - E_o)]. \tag{2.27}$$

Similarly, since Kepler's equation evaluated at times t and t_o yields

$$\left[\frac{\mu}{a^3}\right]^{1/2}(t - t_o) = (E - E_o) - e(\sin E - \sin E_o), \tag{2.28}$$

g may be written as

$$g = (t - t_o) - \left[\frac{a^3}{\mu}\right]^{1/2}[(E - E_o) - \sin(E - E_o)] \tag{2.29}$$

The variables f and g and their time derivatives are now written as functions of the change in eccentric anomaly, which may be found using a modified form of Kepler's equation (see Prob. 2.15.) The position and velocity of the object may thus be found directly as a function of the conditions at epoch, $\mathbf{r}_o$ and $\mathbf{v}_o$, and time elapsed since epoch $t - t_o$. In the process, however, some form of Kepler's equation must be solved by iteration.

A series expansion in the elapsed time parameter $t - t_o$, which is due to Gauss, can also be used to find f and g, and this does away with the need to solve Kepler's equation.

Expanding $\mathbf{r}$ in a Taylor series about $t = t_o$ yields

$$\mathbf{r}(t) = \sum_{n=0}^{\infty} \frac{(t - t_o)^n}{n!}\left[\frac{d^n\mathbf{r}}{dt^n}\right]_{t_o} \tag{2.30}$$

Since the motion occurs in a single plane, all of the derivatives $d^n\mathbf{r}/dt^n$ must lie in the plane of $\mathbf{r}$ and $\mathbf{v}$. Therefore,

$$\frac{d^n\mathbf{r}}{dt^n} = F_n\,\mathbf{r} + G_n\,\mathbf{v} \tag{2.31}$$

so,

$$\frac{d^{n+1}\mathbf{r}}{dt^{n+1}} = \dot{F}_n\,\mathbf{r} + F_n\,\mathbf{v} + \dot{G}_n\,\mathbf{v} + G_n\,\dot{\mathbf{v}} \tag{2.32}$$

Position in Orbit as a Function of Time

but

$$\dot{\mathbf{v}} = \ddot{\mathbf{r}} = \frac{-\mu \mathbf{r}}{r^3} \tag{1.24}$$

and let $h \equiv \mu/r^3$, then

$$\frac{d^{n+1}\mathbf{r}}{dt^{n+1}} = (\dot{F}_n - hG_n)\mathbf{r} + (F_n + \dot{G}_n)\mathbf{v}$$

$$= F_{n+1}\mathbf{r} + G_{n+1}\mathbf{v}, \tag{2.33}$$

by definition, with $F_o = 1$ and $G_o = 0$ from (2.31). Then

$$F_1 = \dot{F}_o - hG_o = 0$$

$$G_1 = F_o + \dot{G}_o = 1$$

$$F_2 = \dot{F}_1 - hG_1 = -h = -\mu/r^3$$

$$G_2 = F_1 + \dot{G}_1 = 0 \tag{2.34}$$

$$F_3 = \dot{F}_2 - hG_2 = -\dot{h} = \frac{3\mu\dot{r}}{r^4} = \frac{3h\dot{r}}{r}$$

$$= 3hp, \text{ where } p \equiv \frac{(\mathbf{r} \cdot \mathbf{v})}{r^2}$$

$$G_3 = F_2 + \dot{G}_2 = -h$$

and F_i and G_i may be obtained to any order using the recursion relation (2.33). Substituting the F_i and G_i into Eqs. (2.31) and (2.30) yields

$$\mathbf{r}(t) = F_o \mathbf{r}(t_o) + (t - t_o)[F_1 \mathbf{r} + G_1 \mathbf{v}]_{t_o}$$

$$+ \frac{(t - t_o)^2}{2!}[F_2 \mathbf{r} + G_2 \mathbf{v}]_{t_o} + \cdots$$

$$= \mathbf{r}_o + (t - t_o)\mathbf{v}_o - \frac{(t - t_o)^2}{2} h_o \mathbf{r}_o$$

$$+ \frac{(t - t_o)^3}{6}[3h_o p_o \mathbf{r}_o - h_o \mathbf{v}_o] + \cdots$$

$$= \mathbf{r}_o[1 - \frac{(t-t_o)^2}{2}h_o + \frac{(t-t_o)^3}{2}h_o p_o + \cdots]$$

$$+ \mathbf{v}_o[(t-t_o) - \frac{(t-t_o)^3}{6}h_o + \cdots]$$

$$= f\mathbf{r}_o + g\mathbf{v}_o \tag{2.35}$$

where $h_o \equiv \mu/r_o^3$ and $p_o \equiv (\mathbf{r}_o \cdot \mathbf{v}_o)/r_o^2$. The series may clearly be extended to any order.

These series approximations to the f and g functions of Eqs. (2.27) and (2.29) are useful because they do not require the iterative solution of some form of Kepler's equation. Unfortunately, the truncation error increases with increasing $t - t_o$ and the radius of convergence of the series is finite, so that the use of the series is limited to moderate values of elapsed time.

2.5 Position versus Time in Hyperbolic and Parabolic Orbits: Universal Variables

The problem of determining position as a function of time for hyperbolic and parabolic orbits can be treated in a manner analogous to that of Secs. 2.2 and 2.3. In each case it is necessary to introduce an auxiliary variable, as eccentric anomaly E was needed for the elliptic case, and to solve indirectly or by successive approximation an equation that takes the place of Kepler's Eq. (2.7). This is described in Roy [2.2] and in Kaplan [2.3]. Rather than discuss these orbit-specific methods here, we will describe a universal formulation due to Battin [2.4] for determining position and velocity as a function of time.

The method is termed *universal* because it may be applied without a priori knowledge of the type of conic section represented by the orbit. It requires only the position $\mathbf{r}_o$ and velocity $\mathbf{v}_o$ of the satellite at an initial time or *epoch* t_o, and it will yield the corresponding quantities for an arbitrary time t. Equation (1.50) yields the semimajor axis:

$$a = \left[\frac{2}{r_o} - \frac{v_o^2}{\mu}\right]^{-1} \tag{2.36}$$

which will be negative for hyperbolic orbits. It is convenient to define

$$\alpha = \frac{1}{a} \tag{2.37}$$

Satellite position and velocity may then be determined as functions of a *universal variable* x whose value is 0 at time t_o:

Position in Orbit as a Function of Time

$$\mathbf{r}(t) = [1 - \frac{x^2}{r_o} C(\alpha x^2)] \mathbf{r}_o$$
$$+ [(t - t_o) - \frac{x^3}{\sqrt{\mu}} S(\alpha x^2)] \mathbf{v}_o \quad (2.38a)$$

$$\mathbf{v}(t) = \frac{x\sqrt{\mu}}{rr_o} [\alpha x^2 S(\alpha x^2) - 1] \mathbf{r}_o + [1 - \frac{x^2}{r} C(\alpha x^2)] \mathbf{v}_o \quad (2.38b)$$

Equations (2.38) are analogous to Eqs. (2.17) and (2.18), the coefficients of $\mathbf{r}_o$ and $\mathbf{v}_o$ in (2.38a) being what were there called the f and g functions. The difference is that f and g are now expressed as functions of the universal variable x, instead of as functions of eccentric anomaly. The value of x to be used in evaluating these coefficients is found from an iterative solution of the *universal Kepler's equation*:

$$\sqrt{\mu}(t - t_o) = \sigma_o x^2 C(\alpha x^2) + (1 - r_o \alpha) x^3 S(\alpha x^2) + r_o x \quad (2.39)$$

where $\sigma_o \equiv \mathbf{r}_o \cdot \mathbf{v}_o / \sqrt{\mu}$.

The functions $C(\ )$ and $S(\ )$ shown in Eqs. (2.38a), (2.38b), and (2.39) are the transcendental functions:

$$C(y) = \frac{1}{2!} - \frac{y}{4!} + \frac{y^2}{6!} - \cdots$$

$$= \frac{1 - \cos\sqrt{y}}{y}, \quad y > 0 \quad (2.40a)$$

$$= \frac{\cosh\sqrt{-y} - 1}{-y}, \quad y < 0$$

and

$$S(y) = \frac{1}{3!} - \frac{y}{5!} + \frac{y^2}{7!} - \cdots$$

$$= \frac{\sqrt{y} - \sin\sqrt{y}}{\sqrt{y^3}}, \quad y > 0 \quad (2.40b)$$

$$= \frac{\sinh\sqrt{-y} - \sqrt{-y}}{\sqrt{-y^3}}, \quad y < 0$$

For the special case of a parabolic orbit, for which $\alpha = 0$, Eqs. (2.38) and (2.39) simplify considerably. Kepler's Eq. (2.39) becomes a cubic polynomial in x, whose one real root is the desired solution.

In the general case Kepler's Eq. (2.39) can be written in the form $F(x) = 0$ with

$$F(x) = \sigma_o \, x^2 \, C\,(\alpha x^2) + (1 - r_o \alpha) \, x^3 \, S\,(\alpha x^2) + r_o \, x - \sqrt{\mu}\,(t - t_o) \quad (2.41)$$

The solution for x is then a zero of the transcendental function $F(x)$.

The derivative of this function with respect to x is simply the magnitude of the radius:

$$F'(x) = r(x) = \sigma_o \, x \, [1 - \alpha x^2 \, S\,(\alpha x^2)] + (1 - r_o \alpha) \, x^2 \, C\,(\alpha x^2) + r_o \quad (2.42)$$

Newton's method may be used for the iterative solution of Eq. (2.39). However, after extensive numerical experiment, a better algorithm for this purpose has been found [Conway (2.5)]. The *Laguerre* algorithm,

$$x_{i+1} = x_i - \frac{n F(x_i)}{F'(x_i) \pm [(n-1)^2 \, (F'(x_i))^2 - n(n-1) \, F(x_i) \, F''(x_i)]^{1/2}} \quad (2.43)$$

applied to the solution of Eq. (2.41) converges to the correct value of x much more rapidly than the Newton method, and, unlike the Newton method, will converge regardless of how poor a starting approximation it is given. The sign ambiguity in the quotient appearing in Eq. (2.43) is resolved by taking the sign of the numerical value of $F'(x_i)$. In addition, the absolute value of the argument of the square root should be taken, as it is possible (for hyperbolic orbits) for this argument to become negative. Whether the Newton or Laguerre method is used to solve Kepler's equation the derivatives

$$\frac{dS(y)}{dy} = \frac{1}{2y} \, [\, C(y) - 3\, S(y) \,] \quad (2.44a)$$

$$\frac{dC(y)}{dy} = \frac{1}{2y} \, [1 - y\, S(y) - 2\, C(y)] \quad (2.44b)$$

prove helpful. The Laguerre method is intended for determining the roots of a polynomial equation of degree n, which of course is not the use to which it is being put here, though it works well regardless. The parameter n is thus somewhat arbitrary; very good results have been obtained using $n = 5$, but the speed of convergence seems relatively insensitive to the choice of n. For example, choosing $n = 4$ or 6 does not provide any consistent improvement in performance. It can be noted that choosing $n = 1$ in Eq. (2.43) yields the ordinary Newton algorithm. For elliptic orbits, the Laguerre method may be used to solve Kepler's equation in the original form (2.7). The rate of convergence for the universal equation is cubic, in comparison to Newton's method, which is quadratic. Convergence is obtained for *any* choice of starting approximation [Conway, (2.5)].

When using universal variables the choice of starting value for the iteration process is made difficult because the variable x lacks the physical significance that the eccentric anomaly has when conventional variables are used. See Prob. 2.15 for the relationship between the universal variable x and the change in eccentric anomaly.

The determination of upper and lower bounds on the solution for x can be helpful in selecting a starting value. As seen in Eq. (2.42) $F'(x) > 0$. Thus any value of x for which $F(x) \geq 0$ is an upper bound on the solution. Similarly, any value of x for which $F(x) < 0$ is a lower bound. Prob. 2.4 explores some of these bounds.

As shown by Prussing [2.6 and 2.8] the solution to Eq. (2.39) lies between upper and lower bounds x^+ and x^- given by

$$x^+ = \frac{\sqrt{\mu}\,(t - t_o)}{r^-} \tag{2.45a}$$

$$x^- = \frac{\sqrt{\mu}\,(t - t_o)}{r^+} \tag{2.45b}$$

where r^+ and r^- are the maximum and minimum values of the radius on the interval $0 \leq x \leq x^+$. For all conic orbits the most conservative upper bound x^+ is obtained by choosing the minimum radius r^- in Eq. (2.43) to be the periapse radius. The most conservative lower bound for elliptic orbits similarly results from choosing the maximum radius r^+ to be the apoapse radius. For hyperbolic orbits the most conservative lower bound is $x^- = 0$, obtained by choosing an infinite value for maximum radius r^+. Several means of obtaining tighter bounds are discussed in Refs. [2.7] and [2.8]. A simple starting value based on the upper and lower bounds is simply the midpoint, that is, $x_o = (x^+ + x^-)/2$.

A more refined universal starting value can be obtained as the secant estimate of the zero of $F(x)$ using the lower bound $x^- = 0$ and the upper bound of Eq. (2.45a) with r^- equal to the periapse radius r_p:

$$x_o = \frac{\mu\,(t - t_o)^2}{r_p\,[F(x^+) + \sqrt{\mu}\,(t - t_o)]} \tag{2.46}$$

Because of the robustness of the Laguerre algorithm all that is required for successful convergence is a reasonable starting value, such as that in Eq. (2.46).

Once the value of the universal variable x has been determined, the corresponding change in true anomaly can be determined directly using the auxiliary variable $z \equiv x/2$ as

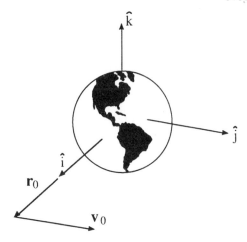

Fig. 2.2   Initial Position and Velocity

$$\tan\left[\frac{f - f_o}{2}\right] = \frac{z\sqrt{p}\ [1 - \alpha z^2 S\ (\alpha z^2)]}{r_o\ [1 - \alpha z^2 C\ (\alpha z^2)] + \sigma_o\ z\ [1 - \alpha z^2 S\ (\alpha z^2)]} \quad (2.47)$$

Example 2.2 [Ref. 2.3]

At a time we will arbitrarily call $t = 0$, as shown in Fig. 2.2, a satellite of the earth has position and velocity vectors with respect to an earth-centered (but inertial) frame given by

$$\mathbf{r}_o = 10{,}000 \text{ km } \hat{i}$$

$$\mathbf{v}_o = 9.2 \text{ km/s } \hat{j}$$

Determine the position $\mathbf{r}$ and velocity $\mathbf{v}$ at time $t = 3$ hours.

From Eq. (2.17) we may solve for the orbit semimajor axis as

$$a = \left[\frac{2}{r_o} - \frac{v_o^2}{\mu}\right]^{-1} = -\,8.104 \times 10^4 \text{ km}$$

so that the orbit must be hyperbolic. Now, $t = 3$ hr $= 10{,}800$ s; $\mathbf{r}_o \cdot \mathbf{v}_o = 0$, so that at time $t = 0$ the satellite must be at perigee. From Eq. (2.37), $\alpha = -\,1.234 \times 10^{-5}$ km^{-1}.

The conservative upper and lower bounds on the solution are

$$x^+ = \frac{\sqrt{\mu}\ t}{r^-} = \frac{\sqrt{\mu}\ t}{r_p} = 682$$

$$x^- = 0 \quad \text{(hyperbolic case)}$$

Table 2.1 Iterative Solutions for Example 2.2

i	x	$S(\alpha x^2)$	$C(\alpha x^2)$	$F(x)$
		Newton Iteration		
0	.3410000E+03	.179041E+00	.562720E+00	.456675E+07
1	.2863135E+03	.175302E+00	.543595E+00	.666744E+06
2	.2752122E+03	.174631E+00	.540176E+00	.229398E+05
3	.2748023E+03	.174607E+00	.540053E+00	.301951E+02
4	.2748018E+03	.174607E+00	.540053E+00	.525267E-04
5	.2748018E+03	.174607E+00	.540053E+00	-.540513E-08
		Laguerre Iteration, n = 5		
0	.3410000E+03	.179041E+00	.562720E+00	.456675E+07
1	.2745277E+03	.174590E+00	.539970E+00	-.152822E+05
2	.2748018E+03	.174607E+00	.540053E+00	.134030E-02
3	.2748018E+03	.174607E+00	.540053E+00	.924047E-09

so that one starting approximation, the midpoint of this interval, would be $x_o = 341$.

Table 2.1 shows the results of iterative solutions of Eq. (2.39) for this example, using the starting value $x_o = 341$, and employing both the Newton iteration method of Eq. (2.15) and the Laguerre iteration method of Eq. (2.43) (with $n = 5$). The Newton iteration requires five iterations to achieve convergence to the solution $x = 274.8$, while the Laguerre method requires only three.

The value of $F'(x)$ used in the iteration algorithm is equal to the radius $r(x)$, and has a value at convergence of 55,804 km. The position and velocity vectors can then be found from Eq. (2.38) to be

$$\mathbf{r} = -30{,}783\,\hat{i} + 46{,}546\,\hat{j} \text{ (km)}$$

$$\mathbf{v} = -3.6149\,\hat{i} + 2.4765\,\hat{j} \quad (\text{km/s})$$

at which point the true anomaly is calculated from Eq. (2.47) to be $f = 123°$.

References

2.1 Danby, J. M. A. and Burkardt, T. M., "The Solution of Kepler's Equation I," *Celestial Mechanics*, **31**, 95–107 (1983).

2.2 Roy, A. E., *Orbital Motion*, Adam Hilger, Ltd., Bristol, 2nd ed. (1982).

2.3 Kaplan. M., *Modern Spacecraft Dynamics and Control*, Wiley & Sons, New York (1976).

2.4 Battin, R. H., *Astronautical Guidance*. McGraw-Hill, New York (1964).

2.5 Conway, B. A., "An Improved Method due to Laguerre for the Solution of Kepler's Equation," *Celestial Mechanics*, **39**, 199–211 (1986).

2.6 Prussing, J. E., "Bounds on the Solution to Kepler's Equation," *J. Astronautical Sciences*, **25**, 123–128 (1977).

2.7 Bergam, M. J. and Prussing, J. E., "Comparisons of Starting Values for Iterative Solution to a Universal Kepler's Equation," *J. Astronautical Sciences* **30**, 75–84 (1982).

2.8 Prussing, J. E., "Bounds on the Solution to a Universal Kepler's Equation," *J. Guidance and Control* **2**, 440–442 (1979).

Problems

2.1 a) Prove that the half-angles $f/2$ and $E/2$ lie in the same quadrant.
b) Determine whether the *principal value* of the function $\tan^{-1}$ used to solve Eq. (2.11) is always correct or whether a quadrant correction is sometimes necessary.

2.2 For low earth orbit (LEO) *approximately* how accurately must E be determined to specify the location of a satellite to within 1 km along the orbit?

2.3 A satellite is in an elliptic orbit with $e = 0.5$. Using a starting value of $E_o = M$, determine the value of E to an accuracy of 10^{-4} radians at the time the satellite is one-quarter period past periapse passage. List all iterations, including the value of $F(E)$ by:
a) using the Newton algorithm, and
b) using the Laguerre algorithm (2.42).

c) Repeat parts a) and b) using the starting value of Eq. (2.16),
d) Calculate the value of f for the solution.
e) Determine, in terms of a the value of r for the solution, and
f) Determine the radial and circumferential components of the velocity vector and its magnitude in units of $\sqrt{\mu/a}$).

2.4* For the function $F(E)$ defined in Eq. (2.13) with $0 \le M \le \pi$, prove that the following conditions that determine bounds on the solution are valid:
a) $F(M) \le 0$,
b) $F(M + e) \ge 0$,
c) $F(M / 1 + e) \le 0$, and
d) $F(M / 1 - e) \ge 0$.
e) What is the condition for which the upper bound of (d) is smaller than the upper bound of (b)?
f) Show that $0 \le E - M \le e$ for $0 \le M \le \pi$ and that $-e \le E - M \le 0$ for $\pi \le M \le 2\pi$.

2.5 For the universal Kepler's Eq. (2.39)
a) Determine a general expression for the values of the universal variable x for which the transcendental function C has value 0.
b) Determine the corresponding time intervals $t - t_o$.
c) Determine the corresponding values of the radius magnitude.

2.6 a) Specialize the universal Kepler's Eq. (2.39) to the case of a parabolic orbit that has $t_o = \tau$, the time of periapse passage.
b)* Derive an expression for the total time interval Δt during which a spacecraft on a parabolic orbit about a planet is inside a sphere of radius $R \ge r_p$ centered at the planet center.
c)* Obtain a simplified approximate expression for Δt which is valid in the case $R \gg r_p$.
d)* Evaluate the expression in part (c) for $R = 144$ earth radii, expressing your answer in days.

2.7 A spacecraft A is in a circular orbit of radius R about a planet. A second spacecraft B is in a circular orbit of radius $R + H$ with $H > 0$.
a) Which spacecraft has the greater orbital period?
b) Determine the approximate difference in orbital period expressed to first order in H.
c) Check your result against the exact difference in periods of two earth satellites in circular orbits of altitudes 300 km and 450 km, respectively.

2.8 A spacecraft is in circular orbit of radius r and speed v about a planet. A motor on the spacecraft is fired, instantaneously increasing the speed by an amount Δv in the direction of motion.
a) What is the eccentricity of the new orbit expressed in terms of $\Delta v/v$?
b) What is the range of values of $\Delta v/v$ for which the new orbit is elliptic?
c) For an elliptic orbit what is the ratio of the apoapse velocity to the original circular orbit speed in terms of $\Delta v/v$.
d) Obtain first-order approximate expressions for (a) and (c) that are valid for small $\Delta v/v$.

2.9 a) Determine for an elliptic orbit an expression for the time required to travel from a point on the minor axis through periapse to the point on the other end of the minor axis.
b) Determine the time to travel from the minor axis through apoapse to the other end of the minor axis.
c) Calculate the sum of the times in (a) and (b) and comment.
d) Calculate the ratio of the time determined in a) to the orbit period. Plot the result as a function of eccentricity for $0 \le e < 1$.

2.10 For an elliptic orbit determine the ratio of the time required to travel from a point on the latus rectum ($f = -\pi/2$) through periapse to the point on the opposite end of the latus rectum ($f = \pi/2$) to the orbit period. Express your answer as a function of eccentricity alone. Check your result in the limit as $e \to 0$.

2.11 An earth-orbiting satellite has a period of 15.743 hr and a perigee radius of 2 earth radii.
a) Determine the semimajor axis of the orbit.
At time $t = 10$ hours after perigee passage, determine:
b) the radius r correct to three significant figures,
c) the velocity magnitude, and
d) the radial component of the velocity.

2.12 In the case of a satellite to be launched into a nominally circular orbit, determine the approximate change in period due to a 1 percent increase in initial radius with no change in initial angular momentum h.

2.13 Canonical length and time units can be defined for which the value of the gravitational constant μ of the central body is unity.
a) For planetocentric orbits a convenient canonical length unit is the planet radius. By considering a unit radius circular orbit, determine the canonical time unit in terms of the period of the circular orbit.

b) Calculate the value in minutes of the canonical time unit for earth (earth radius = 6378 km).

c) Calculate an alternative canonical length unit that, along with the time unit of seconds, will make earth g have a value of unity. Note that this makes mass and weight numerically equal.

d) For heliocentric orbits the earth–sun mean distance is a convenient canonical length unit, defined as 1 astronomical unit (au). Using this, determine the value in days of the time unit for which μ of the sun is equal to unity. Note that another useful unit is obtained by defining the speed in a circular orbit of radius 1 au as 1 EMOS (earth mean orbital speed).

2.14 Show that in Fig. 2.1, when $f = \pi/2$, the length of the line segment FQ is equal to the semiminor axis b.

2.15* Starting with the simple form of Kepler's Eq. (2.7), derive the form valid for an arbitrary epoch t_o:

$$n(t - t_o) = E - E_o + \frac{\sigma_o}{\sqrt{a}} [1 - \cos(E - E_o)] - (1 - \frac{r_o}{a}) \sin(E - E_o)$$

where $\sigma_o \equiv r_o \cdot v_o / \sqrt{\mu}$. Note the similarity to Eq. (2.41) and show that the universal variable x is related to the change in eccentric anomaly for elliptic orbits by $x = \sqrt{a}\ (E - E_o)$.

2.16* Verify the identities given in Eq. (2.44).

2.17* Starting with $F(x)$ in Eq. (2.41), derive $F'(x)$ given in Eq. (2.42) and also $F''(x)$ for use in the Laguerre algorithm.

2.18 Use the f and g functions of Sec. 2.4 to evaluate the position and velocity vectors $\mathbf{r}$ and $\mathbf{v}$ at the apoapse of an elliptic orbit in terms of $\mathbf{r}_o$ and $\mathbf{v}_o$ at periapse ($t_o = \tau$). Verify that your results are correct.

3
The Orbit in Space

3.1 Introduction

Six constants are required to completely specify the orbit of a particular satellite with respect to the attracting center. In the most elementary form the six components of the vectors **r** and **v** at a specified time will serve this purpose. We have seen from the previous section on universal variables that with a knowledge of these two vectors at any initial time or *epoch* t_o, position and velocity at any future time may be determined. Unfortunately, **r** and **v** do not directly yield much information about the orbit. For example, they do not explicitly tell us what type of conic the orbit represents. Another set of six constants, the *orbital elements*, is much more descriptive of the orbit.

3.2 The Orbital Elements

From the fact that the vector **h** = **r** x **v** is a constant (Eq. 1.26) we know that the plane of the orbit is fixed in space. The vectors **r** and **v** lie in this plane and **h** is normal to it. Now let us define a reference plane, also assumed fixed in space. The orbit and reference planes are shown in Fig. 3.1. The first constant we may then define is the *inclination*, i, the angle between the planes. The intersection of the two planes is a line. The direction of this line, called the *line of nodes* will be the second constant, but to specify its direction we need a "reference line" in the same way we needed a reference plane. To describe the size and shape of the orbit we require the semimajor axis, a, and the eccentricity, e, the third and fourth constants. To describe the orientation of the orbit in the plane it is sufficient to locate the position of the periapse; this may be measured from the line of nodes by the angle ω, called the *argument of periapse*. Finally, the position of the satellite along the orbit may be given as before, by the true anomaly f, measured at epoch from the periapse.

The customary choice for the "reference line" is the direction from the sun to a fixed point on the celestial sphere. The chosen point is called the *first point in Aries* and is symbolized by ♈, a stylized ram's head. As seen from the earth, the sun will lie directly "over" this point at the vernal equinox, as seen in Fig. 3.2. The sun would thus be said to be "entering Aries" on this day. Of course, any point on the celestial sphere would do as

The Orbit in Space 47

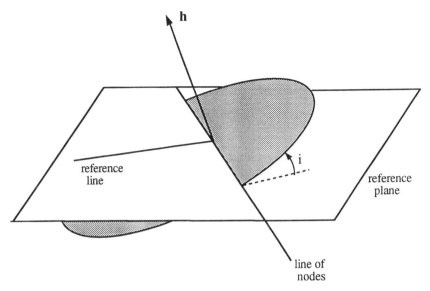

Fig. 3.1   Orbit and Reference Planes

well, if it were universally agreed upon.

The direction of the line of nodes, or actually the position of the ascending node, where the satellite would rise above (i.e., in a northerly direction) the reference plane, is denoted by Ω and called the *longitude of the ascending node*. The constants describing the orbit in space are thus a, e, i, Ω, ω, and f; the angle orbital elements are shown in Fig. 3.3.

The reference plane remains to be chosen. Any space-fixed plane would do, but there are two obvious choices. For a satellite of the earth, the earth's equatorial plane would be a convenient reference, as shown in Fig. 3.4. However, this plane is not space-fixed because the earth's spin axis precesses (describes a cone) with a period of approximately 26,000 years. On short time scales (e.g. a satellite lifetime), the earth's equatorial plane may be considered inertially fixed without introducing much error. However, for very precise determinations of position, it is necessary to refer longitude measurements (see Fig. 3.3) to the position of the reference line (or vernal equinox) of a particular date, or *epoch*. For convenience the epoch is chosen not too distantly in the past nor is it changed often; recent epochs are 1900 and 1950. For interplanetary trajectories, in which the sun is the attracting center, a more convenient reference plane is the *ecliptic plane*, the plane of the earth's orbit about the sun, illustrated in Fig. 3.2. Since the line of intersection of the earth's equatorial plane and the ecliptic plane lies in both planes and locates the first point in Aries, it is a convenient reference line from which to measure longitude, whether one is describing geocentric or

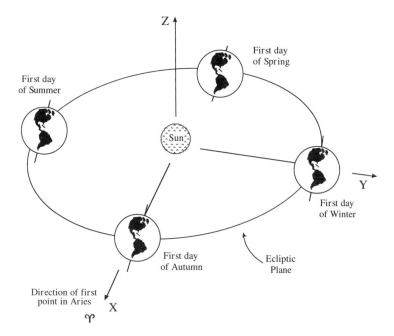

Fig. 3.2 The Position of the Earth with Respect to the Sun at the Beginning of Each Season (Northern Hemisphere).

heliocentric motion. The definitions of Fig. 3.3 apply for either choice of reference plane.

Because of the previously mentioned slow precession of the earth's spin axis, and hence also of the equatorial plane and the equinoxes, the sun is no longer in Aries on the vernal equinox (March 21) as it was centuries ago when the equinox was first located, but is now in the constellation Pisces. For the same reason the star Polaris is only temporarily the "pole star"; the north celestial pole describes a circular path with respect to the background of the fixed stars.

There are minor variations in the definition of this set of orbital elements which will be found. One is the substitution of either mean anomaly at epoch, M or time of periapse passage, τ, for the true anomaly at epoch. This represents no significant change as, for given a and e, true anomaly, and hence position in orbit, can be found at a specified epoch from either mean anomaly or time of periapse passage through the solution of Kepler's equation, as described in Sec. 2.2.

Other possible choices for orbital elements are the *longitude of periapse* $\tilde{\omega}$, defined as $\Omega + \omega$, the *argument of latitude at epoch* θ, defined as $\omega + f$, and the *true longitude at epoch* L, defined as $\Omega + \omega + f$, which is

The Orbit in Space 49

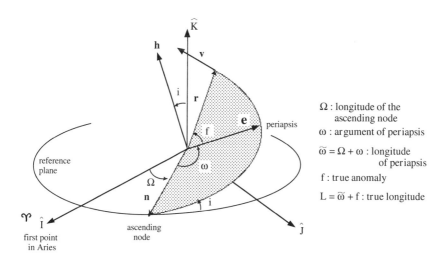

Fig. 3.3 The Angle Orbital Elements

the sum of all the angles from the reference line to the position at epoch. Note that in general these angles are not all in the same plane.

Some of the alternative orbital elements become useful in cases where one or more of the customary elements are undefined, as for example when the inclination or eccentricity is 0. When the inclination is 0, Ω and ω are undefined, but $\tilde{\omega}$ and L are defined. These *singular* cases are considered in Probs. 3.1 and 3.7. In general, six orbital elements are required to uniquely define the orbit. In the singular cases fewer than six elements are needed.

3.3 Determining the Orbital Elements from *r* and *v*

As shown in the previous section, the orbital elements contain the same information as do the position and velocity vectors at a specified time. It must be possible, therefore, to transform from one set of constants to the other. Of course the transformation may be done in either direction; each corresponds to the solution of a practical problem. For orbit determination purposes, radar observations directly yield **r** and may determine **v**. From each observation the orbital elements can be determined by a process which will be described here; and then by using a weighted averaging method the "best fit" orbit can be found. This subject is treated in Chap. 10. For prediction purposes, for example to aim an antenna at a satellite at a given time, the

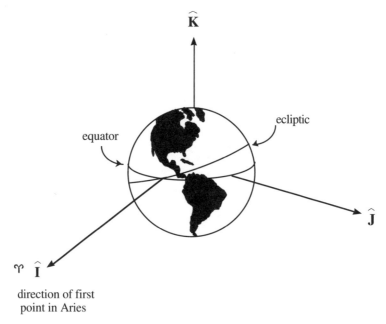

Fig. 3.4 Illustrating the Geocentric-Equatorial Reference Frame (Reference Frame Attached to the Earth with Space-Fixed Basis Vectors)

customary elements are transformed to yield the position vector **r**. Simple geometrical relationships then yield the point on the celestial sphere at which the antenna should be aimed.

For both purposes, orbit determination and position prediction, it is useful to have a commonly accepted set of coordinates for describing the position of a body on the celestial sphere (the background of the "fixed" stars). When it is convenient to use the earth's equator as a reference (e.g., for the case of an earth-orbiting object viewed from the earth) the *declination*, δ, and *right ascension*, α, illustrated in Fig. 3.5, are often used. The reference $(\hat{\mathbf{I}} - \hat{\mathbf{J}})$ plane is the earth's equatorial plane and right ascension is measured positive eastward from the first point in Aries.

Where it is convenient to use the ecliptic plane as a reference (e.g., for the case of an interplanetary trajectory) the *celestial longitude*, λ, and *celestial latitude*, β, may be used. λ and β are analogous to α and δ, respectively, except that the reference plane is the ecliptic plane. Celestial longitude λ is measured positive eastward from the first point in Aries (i.e., from the direction of $\hat{\mathbf{I}}$), which lies in both the ecliptic and equatorial planes. The relationship between the two sets of angles is given by:

The Orbit in Space

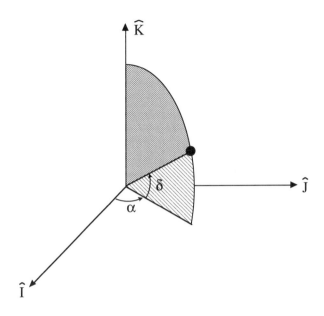

Fig. 3.5 Right Ascension and Declination

$$\cos \delta \cos \alpha = \cos \beta \cos \lambda$$

$$\cos \delta \sin \alpha = \cos \beta \sin \lambda \cos \varepsilon - \sin \beta \sin \varepsilon \qquad (3.1)$$

$$\sin \delta = \cos \beta \sin \lambda \sin \varepsilon + \sin \beta \cos \varepsilon$$

where ε is the *obliquity of the ecliptic*, the angle between the earth's equatorial plane and the ecliptic, equal to 23° 27′.

A knowledge of (α, δ) or (λ, β) from observation of an object clearly determines the unit vector $\mathbf{r}/r$ and vice versa.

Now, assuming that $\mathbf{r}$ and $\mathbf{v}$ are known for an orbiting object at some instant, the semimajor axis may be determined directly from the vis-viva equation (1.50):

$$a = \frac{r}{2 - \frac{rv^2}{\mu}}$$

Recall that a vector constant of integration was introduced in the solution of the equation for the two-body problem, which appeared in the solution of the equation

$$\ddot{\mathbf{r}} \times \mathbf{h} = \mu \frac{d}{dt}\left(\frac{\mathbf{r}}{r}\right) \tag{1.29}$$

which was integrated to yield

$$\dot{\mathbf{r}} \times \mathbf{h} = \frac{\mu \mathbf{r}}{r} + \mu \mathbf{e} \tag{1.30}$$

where $\mathbf{h} = \mathbf{r} \times \mathbf{v}$ is a vector normal to the orbit plane. The vector $\mathbf{e}$ is a constant vector; we later found out that it points toward the periapse from the attracting center and that its magnitude e represents the orbit eccentricity. From Eq. (1.30) we have

$$\mu \mathbf{e} = \mathbf{v} \times \mathbf{h} - \frac{\mu \mathbf{r}}{r}$$

$$= [\mathbf{v} \times (\mathbf{r} \times \mathbf{v}) - \frac{\mu \mathbf{r}}{r}]$$

$$= [(\mathbf{v} \cdot \mathbf{v})\mathbf{r} - (\mathbf{r} \cdot \mathbf{v})\mathbf{v}] - \frac{\mu \mathbf{r}}{r}$$

$$= [(v^2 - \frac{\mu}{r})\mathbf{r} - (\mathbf{r} \cdot \mathbf{v})\mathbf{v}] \tag{3.2}$$

from which the eccentricity e may be determined given $\mathbf{r}$ and $\mathbf{v}$.

Now define the *nodal vector* $\mathbf{n}$ by

$$\mathbf{n} = \hat{\mathbf{K}} \times \frac{\mathbf{h}}{h} \tag{3.3}$$

where $\hat{\mathbf{K}}$ is the unit vector normal to the (earth's) equatorial plane, as shown in Fig. 3.4. The magnitude of the nodal vector $n = \sin i$. Then, since $\mathbf{h}$ is normal to the orbit plane, $\mathbf{n}$ must lie in both the orbit and equatorial planes and hence along the line of nodes. In fact, with reference to Figs. 3.1 and 3.4 it can be seen that $\mathbf{n}$ points toward the ascending node. Then, by inspection,

$$\cos i = \frac{\mathbf{h} \cdot \hat{\mathbf{K}}}{h} = \frac{h_z}{h} \tag{3.4}$$

yields the orbit inclination in the correct range $0 \leq i \leq \pi$. Similarly,

$$\cos \Omega = \frac{\mathbf{n} \cdot \hat{\mathbf{I}}}{n} = \frac{n_x}{n} \tag{3.5}$$

where $0 \leq \Omega \leq \pi$ if $\mathbf{n} \cdot \hat{\mathbf{J}} = n_y \geq 0$ (see Fig. 3.3). Thus the *principal value* of the function $\cos^{-1}$ used to solve Eq. (3.5) yields the correct value. Recall that the principal value of an inverse trigonometric function is the value your calculator or a computer provides. *WARNING*: If $n_y < 0$, then Ω is in the range $\pi < \Omega < 2\pi$ and *the principal value is incorrect. One must then subtract the principal value of $\cos^{-1}$ used to solve Eq. (3.5) from 2π.*

By the definition of $\mathbf{e}$, the argument of periapse is given by

$$\cos \omega = \frac{\mathbf{n} \cdot \mathbf{e}}{ne} \tag{3.6}$$

Here $0 \leq \omega \leq \pi$ if $\mathbf{e} \cdot \hat{\mathbf{K}} = e_z \geq 0$, but if $e_z < 0$, then $\pi < \omega < 2\pi$.

The true anomaly may be found from

$$\cos f = \frac{\mathbf{e} \cdot \mathbf{r}}{er} \tag{3.7}$$

where $0 \leq f \leq \pi$ if $\mathbf{r} \cdot \mathbf{v} \geq 0$, but if $\mathbf{r} \cdot \mathbf{v} < 0$, then $\pi < f < 2\pi$.

The inverse problem, finding $\mathbf{r}$ and $\mathbf{v}$ from the orbital elements, is less straightforward. One method of several possible methods will be presented here.

1. Solve for $r = \dfrac{a(1-e^2)}{1+e \cos f}$

2. Find the unit vector $\hat{\mathbf{r}}$ in the direction of $\mathbf{r}$ by rotating the unit vector $\hat{\mathbf{I}}$ through the Euler angles Ω, i, and θ in succession. See Ref. [3.1] as well as Eq. (9.35).

3. Solve for the magnitude of the velocity, using Eq. (1.50),

$$v^2 = \mu\left(\frac{2}{r} - \frac{1}{a}\right)$$

4. Solve for the flight path angle γ from

$$h = rv \cos \gamma = [\mu a (1 - e^2)]^{\frac{1}{2}}$$

5. Define a vector normal to the orbit plane by

$$\mathbf{c} = \mathbf{r} \times \hat{\mathbf{n}} = \mathbf{r} \times (\cos \Omega \, \hat{\mathbf{I}} + \sin \Omega \, \hat{\mathbf{J}})$$

6. The three components of **v** may then be found by solving the three independent equations:

$$v^2 = v_x^2 + v_y^2 + v_z^2$$

$$\mathbf{r} \cdot \mathbf{v} = rv \sin \gamma = r_x v_x + r_y v_y + r_z v_z$$

$$\mathbf{c} \cdot \mathbf{v} = 0 = c_x v_x + c_y v_y + c_z v_z$$

where $\sin \gamma > 0$ if the value of f represents a point on the outbound part of the conic orbit and $\sin \gamma < 0$ for a point on the inbound part.

An alternative method, which yields explicit expressions for **r** and **v** as functions of the orbital elements is to solve for **r** as in steps (1) and (2), yielding,

$$\mathbf{r} = r \ (\cos \Omega \cos \theta - \sin \Omega \sin \theta \cos i \ ) \ \hat{\mathbf{I}}$$

$$+ r \ (\sin \Omega \cos \theta + \cos \Omega \sin \theta \cos i \ ) \ \hat{\mathbf{J}}$$

$$+ r \sin \theta \sin i \ \hat{\mathbf{K}} \tag{3.8}$$

Then **v** may be found as $\dot{\mathbf{r}}$, or

$$\mathbf{v} = -\frac{\mu}{h} [\cos \Omega \ (\sin \theta + e \sin \omega) + \sin \Omega \ (\cos \theta + e \cos \omega) \cos i \ ] \ \hat{\mathbf{I}}$$

$$+ \frac{\mu}{h} [\sin \Omega \ (\sin \theta + e \sin \omega) - \cos \Omega \ (\cos \theta + e \cos \omega) \cos i \ ] \ \hat{\mathbf{J}}$$

$$+ \frac{\mu}{h} (\cos \theta + e \cos \omega) \sin i \ \hat{\mathbf{K}} \tag{3.9}$$

3.4 Velocity Hodographs

As shown in Fig. 3.6, consider the unit vectors $\mathbf{u}_e$ and $\mathbf{u}_{ne}$, along and normal to the eccentricity vector **e**, with $\mathbf{u}_{ne} = \mathbf{u}_h \times \mathbf{u}_e$. Also consider the *rotating* unit vectors $\mathbf{u}_r$ and $\mathbf{u}_{nr}$, along and normal to the radius vector **r**, with $\mathbf{u}_{nr} = \mathbf{u}_h \times \mathbf{u}_r$. The left-hand side of Eq. (3.10) can be simplified as follows:

$$\mathbf{h} \times (\mathbf{v} \times \mathbf{h}) = \mathbf{v} \ (\mathbf{h} \cdot \mathbf{h}) - \mathbf{h} \ (\mathbf{h} \cdot \mathbf{v}) = h^2 \mathbf{v} \tag{3.11}$$

The Orbit in Space

The right-hand side of Eq. (3.10) can be written as:

$$h\ \mathbf{u}_h \times \mu\ (\mathbf{u}_r + e\ \mathbf{u}_e) = \mu h\ \mathbf{u}_{nr} + \mu e h\ \mathbf{u}_{ne} \tag{3.12}$$

using the unit vectors defined in Fig. 3.6. Combining Eqs. (3.11) and (3.12) yields:

$$\mathbf{v} = \frac{\mu}{h}\ \mathbf{u}_{nr} + \frac{\mu e}{h}\ \mathbf{u}_{ne} \tag{3.13}$$

One then recognizes the surprising result that the components of the velocity vector on an arbitrary conic orbit are *constants* when expressed in terms of the skewed unit vectors $\mathbf{u}_{nr}$ and $\mathbf{u}_{ne}$.

In order to express the components of velocity in an orthogonal set of unit vectors, note that (Fig. 3.6):

$$\mathbf{u}_{ne} = \mathbf{u}_r\ \sin f + \mathbf{u}_{nr}\ \cos f \tag{3.14}$$

Substituting into Eq. (3.13) and multiplying by h/μ to nondimensionalize the result yields:

$$\frac{h}{\mu}\ \mathbf{v} = e\ \sin f\ \mathbf{u}_r + (1 + e\ \cos f)\ \mathbf{u}_{nr} \tag{3.15}$$

As shown in Figs. 3.7 through 3.9, the components given in Eq. (3.15) of the nondimensional velocity vector $(h/\mu)\ \mathbf{v}$ along the *rotating* $\mathbf{u}_r$, $\mathbf{u}_{nr}$ axes *describe a circle of radius e centered at (0, 1)*. This circle is then the

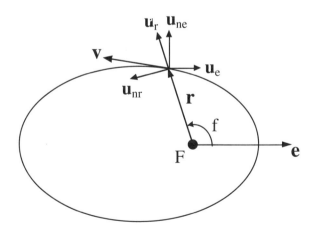

Fig. 3.6 Unit Vector Definitions

velocity hodograph for a conic orbit. Hodographs such as this provide a graphical description of the behavior of vector quantities during the motion. This can be useful both qualitatively and quantitatively. For example, it is clear from Fig. 3.7 that the radial component of velocity is maximum at $f = \pi/2$. The hodographs can also be used for approximate graphical solutions using a protracter and ruler. For example, one can determine the ratio of speed at $f = 120°$ to periapse speed very quickly this way.

In Fig. 3.7 representing an elliptic orbit, the radius of the circle e is less than unity. The flight path angle γ is shown as the angle between the velocity vector and the local horizontal direction, given by $\mathbf{u}_{nr}$. The $\mathbf{u}_{nr}$ component of nondimensional velocity, $1 + e \cos f$, is simply p/r by Eq. (1.35), as shown. Also shown is the eccentricity vector $\mathbf{e}$, which is tangent to the circle with magnitude equal to the radius of the circle. The point on the circle for which the tangent intersects the origin provides a graphical description of the semimajor axis a.

In Fig. (3.8) for a parabolic orbit, the circle intersects the origin because $e = 1$. Because of this it is readily evident that $\gamma = f/2$ and that $\gamma \to \pi/2$ and $v \to 0$ as $r \to \infty$.

As shown in Fig. 3.9 even more information can be displayed for a hyperbolic orbit. The additional variables are the nondimensional inbound and outbound hyperbolic excess vectors $(h/\mu) \mathbf{v}_{\infty_i}$ and $(h/\mu) \mathbf{v}_{\infty_o}$ as well as the limiting value of true anomaly f_∞ for which $f \to f_\infty$ as $r \to \infty$, the turn angle δ of the hyperbola (the angle between the excess velocity vectors in an inertial frame), and the aiming radius Δ (the distance from the focus to the asymptote measured normal to the asymptote). Both δ and Δ were defined previously in Fig. 1.6. What does the portion of the circle below the horizontal axis in Fig. 3.9 represent?

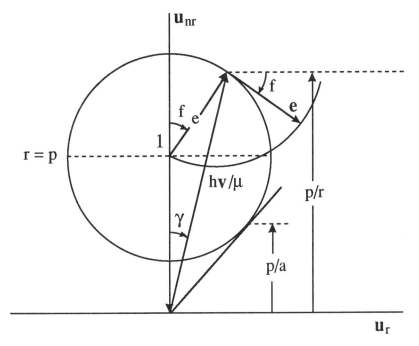

Fig. 3.7 Velocity Hodograph for an Elliptic Orbit

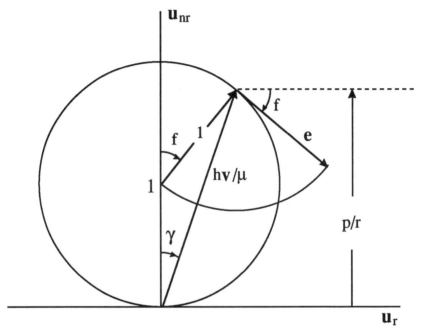

Fig. 3.8 Velocity Hodograph for a Parabolic Orbit

The Orbit in Space 59

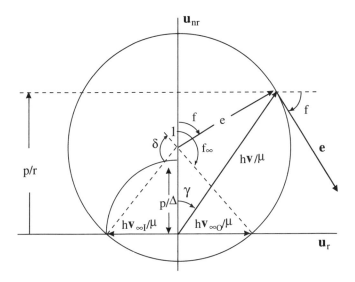

Fig. 3.9 Velocity Hodograph for a Hyperbolic Orbit

Reference

3.1 Greenwood, D. T., *Principles of Dynamics*, Prentice-Hall, 2nd ed. (1988), Chap. 7.

Problems

3.1 For the special cases
  a) $e = 0$ and $\sin i \neq 0$,
  b) $e \neq 0$ and $\sin i = 0$, and
  c) $e = \sin i = 0$, form two lists using the possible orbital elements: one of those that are defined and one of those that are not. Give the

range of values for those that are defined. In each case, provide a complete set of independent elements, showing how each is calculated from **r** and **v**.

3.2 In canonical units ($\mu = 1$) the following data is given. Determine a set of independent orbital elements for each case.

a)
$$\mathbf{r} = -\frac{3\sqrt{2}}{2}\hat{\mathbf{I}} - \frac{3\sqrt{2}}{2}\hat{\mathbf{K}}$$

$$\mathbf{v} = -\frac{\sqrt{6}}{6}\hat{\mathbf{J}}$$

b)
$$\mathbf{r} = \frac{\sqrt{2}}{2}\hat{\mathbf{I}} + \frac{\sqrt{2}}{2}\hat{\mathbf{J}}$$

$$\mathbf{v} = -\frac{\sqrt{2}}{2}\hat{\mathbf{I}} + \frac{\sqrt{2}}{2}\hat{\mathbf{J}}$$

3.3 Determine and sketch the velocity hodographs for conic orbits in terms of the unit vectors $\mathbf{u}_e$ and $\mathbf{u}_{ne}$. Comment on this formulation compared to using the unit vectors $\mathbf{u}_r$ and $\mathbf{u}_{nr}$.

3.4 A satellite in orbit about the earth has instantaneous position $\mathbf{r} = 6045\,\hat{\mathbf{I}} + 3490\,\hat{\mathbf{J}}$ km and velocity specified as $\mathbf{v} = -2.457\,\hat{\mathbf{I}} + 6.618\,\hat{\mathbf{J}} + 2.533\,\hat{\mathbf{K}}$ km/s. Determine the orbital elements. (The earth's equatorial plane is the reference plane, as in Fig. 3.4).

3.5 At what point in an orbit is the declination equal to the orbit inclination?

3.6 The orbital elements of an earth satellite are $a = 1.1$ earth radii, $e = .05$, $i = 45°$, $\Omega = 0°$, $\omega = 20°$, and $f = 10°$. Determine the instantaneous right ascension α and declination δ of the satellite. Use the earth's equatorial plane as the reference plane, as in Figs. 3.4 and 3.5.

3.7* Investigate the use of so-called equinoctial orbital elements to avoid the singularities in the classical elements addressed in Prob. 3.1. The equinoctial elements are $P_1 = e \sin \tilde{\omega}$, $P_2 = e \cos \tilde{\omega}$, $Q_1 = \tan \frac{1}{2}i \sin \Omega$, and $Q_2 = \tan \frac{1}{2}i \cos \Omega$. These replace the classical

elements e, i, Ω, and ω.
a) Do these elements circumvent all the singularities encountered in Prob. 3.1?
b) Derive the relationships that express the classical elements in terms of the equinoctial elements.

3.8 If the time of periapse passage τ is desired, rather than the true anomaly at epoch f given by Eq. (3.7), outline a procedure for calculating τ from f.

3.9 Using the velocity hodograph for an elliptic orbit of eccentricity e equal to 0.5, graphically determine the following quantities and verify their values using the appropriate formulas.
a) The value of the flight path angle γ at $f = 45°$.
b) The minimum value of the flight path angle on the orbit.
c) The value of the radius r for part (a) in terms of the parameter p.
d) The value of r for part (b).
e) Will specification of the value of semimajor axis a determine numerical values to parts (c) and (d)?

3.10 Using the velocity hodograph for a hyperbolic orbit of eccentricity e equal to 2, graphically determine the following quantities and verify their values using appropriate formulas.
a) The turn angle δ.
b) The limiting value of true anomaly f_∞.
c) The value of hyperbolic excess speed v_∞ in units of μ/h.
d) The aiming radius Δ in terms of the parameter p.
e) For what eccentricity value is the aiming radius of a hyperbola equal to the parameter p?

3.11 Using the velocity hodograph for a parabolic orbit, graphically determine the limiting value of the vector $\mathbf{e}$ as $r \to \infty$.

4
Lambert's Problem

4.1 Introduction

A fundamental problem in astrodynamics is the transfer of a spacecraft from one point in space to another. An example application is spacecraft targeting, in which the final point (the "target") is a planet or space station moving in a known orbit. In this situation, one might want the spacecraft to either *intercept* the target (match position only) or *rendezvous* with the target (match both position and velocity).

The initial point for an orbital rendezvous or interception is typically the location of the spacecraft in its orbit at the initial time. However, in other applications, such as ascent trajectories from the surface of the moon, the initial point can be at rest on the surface. Common to all orbit transfer applications is the determination of two-body orbits that connect specified initial and final points.

4.2 Transfer Orbits Between Specified Points

As shown in Fig. 4.1, consider points P_1 and P_2 described by radius vectors $\mathbf{r}_1$ and $\mathbf{r}_2$ relative to the focus F at the center of attraction. The end points P_1 and P_2 are separated by the transfer angle θ and the chord c. The triangle FP_1P_2 is sometimes referred to as the *space triangle* for the transfer.

First, let us investigate the possible transfer orbits between the specified endpoints P_1 and P_2. For the case of elliptic transfer orbits, this can be accomplished using the simple geometric property shown in Fig. 4.2. A similar analysis can be done for parabolic and hyperbolic transfer orbits, but is not presented here.

The geometric property used is that the sum of the distances from any point on the ellipse to the focus and the vacant focus is a constant having value $2a$. This is the familiar property by which one can draw an ellipse by anchoring a piece of string at two points with thumbtacks, draw the string taut with the point of a pencil, and trace out an ellipse. In this mechanical device the thumbtacks locate the focus and vacant focus, and the string length is $2a$. Thus in Fig. 4.2

Lambert's Problem

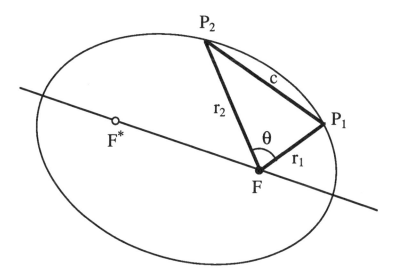

Fig. 4.1 Transfer Orbit Geometry

$$P_1F + P_1F^* = 2a \qquad (4.1)$$

and

$$P_2F + P_2F^* = 2a$$

or

$$P_1F^* = 2a - r_1 \qquad (4.2)$$

and

$$P_2F^* = 2a - r_2$$

For the remainder of the discussion, let us assume that $r_2 \geq r_1$, which implies no loss of generality, since the transfer orbit can be traversed in either direction. Because gravity is a conservative (nondissipative) force, one can determine the orbit that solves the boundary value problem in the reverse direction (starting at the final point P_2 and ending at the initial point P_1) by simply letting time run backward on the original orbit from P_2 to P_1. This represents a valid forward time solution with the original velocity vector replaced by its negative.

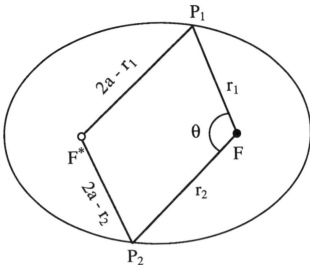

Fig. 4.2 A Geometric Property of Ellipses

For a given space triangle and specified value of semimajor axis a, Fig. 4.3 shows that a vacant focus is located at the intersection of two circles centered at P_1 and P_2 having respective radii $2a - r_1$ and $2a - r_2$ [Eq. (4.2)].

As shown in Fig. 4.3, the circular arcs for a given value of $a = a_k$ intersect at two points labeled F_k^* and $\tilde{F}_k^*$ that are equidistant from the chord c. This means that for the value of a depicted, there are two elliptic transfer orbits between P_1 and P_2. As we will see, these two transfer orbits for the same value of a have different eccentricities and transfer times, but they have the same total energy.

From Fig. 4.3 it is evident that the distance FF^* is less than the distance $F\tilde{F}^*$. Because the distance from the focus of an ellipse to the vacant focus is $2ae$ (Sec. 1.5), this implies that the ellipse with vacant focus at F^* has the smaller eccentricity: $e < \tilde{e}$. Figure 4.4 shows the two elliptic transfer orbits for the case $r_2 = 1.524\, r_1$ (earth to Mars) with $\theta = 107°$ and a specified semimajor axis value of $a = 1.36\, r_1$. The numerical values of the two eccentricities are $e = 0.26$ and $\tilde{e} = 0.68$ for this case.

Returning to Fig. 4.3 two other aspects of the problem are evident from the geometry. First, as the value of a is varied, the vacant foci describe a locus formed by the intersections of the circles of varying radii centered at P_1 and P_2. This *locus of the focus* has the property that at any point on it the *difference* in the distances to the fixed points P_1 and P_2 is a constant, equal to $r_2 - r_1$. This implies that the locus itself (the solid line in Fig. 4.3) is a *hyperbola* with foci at P_1 and P_2 !

Lambert's Problem

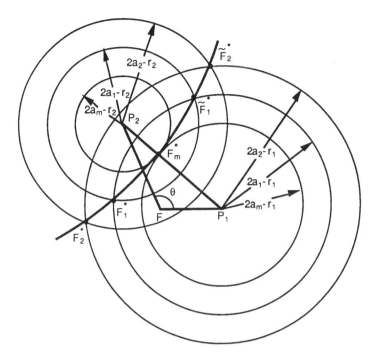

Fig. 4.3 Vacant Focus Locations

The second aspect of the problem that is evident from Fig. 4.3 is that as the value of a is decreased from the values shown, the two vacant foci approach the point F_m^* on the chord between P_1 and P_2. For all values of a less than this value there is *no intersection* of the circles centered at P_1 and P_2, which implies that no elliptic transfer connecting P_1 and P_2 exists for values of a less than a certain minimum value. This minimum value is denoted by a_m and its value is easily calculated from the geometry of the point F_m^* in Fig. 4.3:

$$(2\,a_m - r_2) + (2\,a_m - r_1) = c \tag{4.3}$$

or

$$a_m = s/2 \tag{4.4}$$

where

$$s \equiv (r_1 + r_2 + c)/2 \tag{4.5}$$

Earth - Mars Transfer

$$r_2 = 1.524\, r_1$$
$$.26 = e < e^* = .68$$
$$\theta = 107°$$
$$a = 1.36\, r_1$$
$$a_m = 1.14\, r_1$$

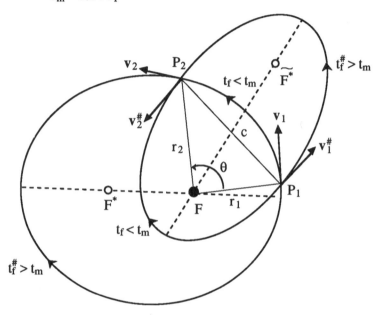

Fig. 4.4 Two Elliptic Transfer Orbits with the Same Value of a

is the *semiperimeter* of the space triangle FP_1P_2.

In addition to describing the value of a_m geometrically, one can also interpret it dynamically by recalling that the value of a for a conic orbit is a measure of its total energy (Sec. 2.4). Thus the ellipse having semimajor axis a_m is the *minimum-energy ellipse* that connects the specified endpoints P_1 and P_2. Orbits having a value of a less than a_m simply do not have enough energy at point P_1 to reach point P_2. Some days are like that.

One other interesting geometric property of the elliptic transfer orbits between P_1 and P_2 concerns the eccentricities of these orbits. As will be shown, the locus of the eccentricity vectors is a straight line that is normal to the chord, as shown by Battin, Fill, and Shepperd in [4.1].

Lambert's Problem

To demonstrate this, one uses the basic polar equation for a conic section

$$r = \frac{p}{1 + e \cos f} \tag{4.6}$$

to write

$$\mathbf{e} \cdot \mathbf{r}_1 = p - r_1 \; ; \; \mathbf{e} \cdot \mathbf{r}_2 = p - r_2 \tag{4.7}$$

Subtracting and dividing by the chord c yields

$$-\mathbf{e} \cdot (\mathbf{r}_2 - \mathbf{r}_1)/c = (r_2 - r_1)/c \tag{4.8}$$

Because $(\mathbf{r}_2 - \mathbf{r}_1)/c$ is a unit vector along the chord directed from P_1 to P_2, Eq. (4.8) implies that the eccentricity vectors for all the transfer orbits have a constant projection along the chord direction. This, in turn, implies that the locus of the eccentricity vectors is a straight line normal to the chord as shown in Fig. 4.5.

Also evident from Fig. 4.5 is the fact that there is a transfer ellipse of *minimum eccentricity* e_s whose value is simply [4.3]

$$e_s = \frac{r_2 - r_1}{c} \tag{4.9}$$

which is, interestingly, the reciprocal of the eccentricity of the hyperbolic locus of the vacant focus. This minimum eccentricity ellipse is also termed the *fundamental ellipse* because the point P_1 has the same relationship to the occupied focus F as P_2 does to the vacant focus $F,^*$ due to the fact that the major axis of the ellipse is parallel to the chord.

4.3 Lambert's Theorem

A primary concern in orbit transfer is the transfer time, defined as the time required to travel from point P_1 to point P_2. In the spacecraft targeting example mentioned earlier, the spacecraft is at point P_1 in its orbit at a time t_1 and the target will be at point P_2 in its orbit at a later time t_2. The transfer time is then $t_2 - t_1$ and the crucial issue is the determination of the transfer orbit that connects the specified endpoints in the given transfer time. Because of the work of Johann Heinrich Lambert (1728–1779), this is often called *Lambert's problem*.

The theorem which bears his name is due to his conjecture in 1761, based on geometric reasoning, that the time required to traverse an elliptic arc between specified endpoints *depends only on the semimajor axis of the ellipse, and on two geometric properties of the space triangle, namely the chord length and the sum of the radii from the focus to points P_1 and P_2*:

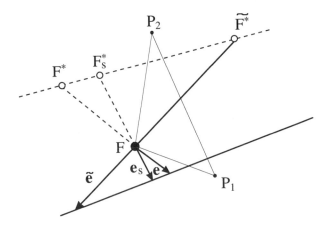

Fig. 4.5 Locus of Eccentricity Vectors

$$t_2 - t_1 = f(a, c, r_1 + r_2) \tag{4.10}$$

This result was actually first obtained analytically by Lagrange in 1778, one year before Lambert's death.

This is the third instance encountered of an important property of elliptic orbits that depends on the semimajor axis but is independent of the eccentricity. The other two are the period of an elliptic orbit and the total energy of the orbit (i.e. the velocity as a function of the radius given by the vis–viva equation).

In order to derive an equation describing the transfer time, we make use of our previously obtained equation relating time and position, namely Kepler's equation. The essential difference here is that Lambert's problem is an orbital *boundary value problem*, whereas Kepler's equation describes an *initial value problem*. If one lets E_2 and E_1 denote the (unknown) values of eccentric anomaly at times t_2 and t_1 on the transfer ellipse, Kepler's equation yields:

$$\sqrt{\mu}\,(t_2 - t_1) = a^{3/2}\,[E_2 - E_1 - e\,(\sin E_2 - \sin E_1)] \tag{4.11}$$

This form is not very convenient for the boundary value problem because both a and e of the transfer orbit are unknown (to be determined) and because, according to Lambert's theorem, the transfer time does not actually depend on the value of e. In order to get the transfer time equation into a

Lambert's Problem

more convenient form, define

$$E_P = \tfrac{1}{2}(E_2 + E_1)$$

and

$$E_M = \tfrac{1}{2}(E_2 - E_1) > 0 \qquad (4.12)$$

Using the fact that $r = a(1 - e \cos E)$,

$$r_1 + r_2 = a[2 - e(\cos E_1 + \cos E_2)] \qquad (4.13)$$

$$= 2a[1 - e \cos E_P \cos E_M] \qquad (4.14)$$

In terms of cartesian coordinates with origin at the geometric center of the ellipse with the x-axis along the major axis (see Fig. 2.1):

$$x = a \cos E \qquad (4.15)$$

$$y = b \sin E \; ; \quad b = a(1 - e^2)^{1/2} \qquad (4.16)$$

and the chord distance can be obtained from

$$c^2 = (x_2 - x_1)^2 + (y_2 - y_1)^2$$

$$= a^2[(\cos E_2 - \cos E_1)^2 + (1 - e^2)(\sin E_2 - \sin E_1)^2]$$

$$= 4a^2 \sin^2 E_M (1 - e^2 \cos^2 E_P) \qquad (4.17)$$

The temptation is irresistible to let

$$\cos \xi = e \cos E_P \qquad (4.18)$$

which is allowable only because the numerical value of e does not exceed unity. This leads to a perfect square on the right-hand side of Eq. (4.17), resulting in

$$c = 2a \sin E_M \sin \xi \qquad (4.19)$$

Equation (4.14) can then be rewritten as

$$r_1 + r_2 = 2a(1 - \cos E_M \cos \xi) \qquad (4.20)$$

Finally, let

$$\alpha = \xi + E_M \qquad (4.21a)$$

$$\beta = \xi - E_M \tag{4.21b}$$

Now one can combine Eqs. (4.19), (4.20), and (4.21) to write:

$$r_1 + r_2 + c = 2a\,(1 - \cos\alpha) = 4a\,\sin^2(\alpha/2) \tag{4.22}$$

$$r_1 + r_2 - c = 2a\,(1 - \cos\beta) = 4a\,\sin^2(\beta/2) \tag{4.23}$$

Equation (4.11) for the transfer time becomes

$$\sqrt{\mu}\,(t_2 - t_1) = 2a^{3/2}\,(E_M - \cos\xi \sin E_M) \tag{4.24}$$

whence our final result is

$$\sqrt{\mu}\,(t_2 - t_1) = a^{3/2}\,[\alpha - \beta - (\sin\alpha - \sin\beta)] \tag{4.25}$$

To summarize, Eq. (4.25), sometimes called *Lambert's equation*, describes the elliptic transfer time $t_2 - t_1$ for the case of less than one complete revolution $0 \le \theta < 2\pi$, where the variables α and β are determined from Eqs. (4.22) and (4.23) as

$$\sin\left[\frac{\alpha}{2}\right] = \left[\frac{s}{2a}\right]^{1/2} \tag{4.26}$$

$$\sin\left[\frac{\beta}{2}\right] = \left[\frac{s-c}{2a}\right]^{1/2} \tag{4.27}$$

As defined before, s is the semiperimeter of the space triangle, equal to $\frac{1}{2}(r_1 + r_2 + c)$. Note that Lambert's theorem as stated in Eq. (4.10) has been proved, since the angles α and β depend only on a, c, and $r_1 + r_2$.

4.4 Properties of the Solutions to Lambert's Equation

Equation (4.25) must provide the solutions for the transfer times on all elliptic arcs of less than one revolution that connect points P_1 and P_2 for a given space triangle and a specified value for a. Returning to Fig. 4.4, it can be seen that for $a > a_m$ there are four such arcs. (Recall that the direction of traversal on these arcs is irrelevant to the analysis because time-reversed solutions are valid as forward time solutions.) Two of these arcs have a transfer angle $\theta < \pi$ as shown; the other two correspond to a transfer angle greater than π, and are formed by the remaining portions of the ellipses containing the smaller transfer angle arcs.

These four solutions for the transfer time correspond to quadrant ambiguities associated with the angles α and β. The principal values of the inverse sine function used to solve Eqs. (4.26) and (4.27) yield angles α_0 and

Lambert's Problem

β_0 characterized by $0 \le \beta_0 \le \alpha_0 \le \pi$. In order to determine which one of the four arcs corresponds to these principal values and what quadrant corrections are needed for the other arcs, it is convenient to utilize the geometric interpretation of the angles α and β derived by Prussing [4.2].

The derivation of the geometric interpretation is based on two properties of elliptic motion: (1) the transfer time must satisfy Kepler's equation, and (2) the shape of the transfer orbit can be altered by moving the focus F and the vacant focus F^* without altering the transfer time or the angles α and β *as long as* $r_1 + r_2$, c, and a remain unchanged. Using this property the focus and vacant focus can be moved to the locations F_R and F_R^* shown in Fig. 4.6, which define a *rectilinear elliptic orbit* ($e = 1$, $p = 0$) between points P_1 and P_2. This rectilinear orbit has the same values of $r_1 + r_2$, c, and a and hence the same transfer time, α and β as the original orbit.

Kepler's equation for the transfer time between two points in an elliptic orbit whose locations are specified by the values of eccentric anomaly E is

$$\sqrt{\mu}\,(t_2 - t_1) = a^{3/2}\,[E_2 - E_1 - e\,(\sin E_2 - \sin E_1)] \qquad (4.28)$$

By comparing Eq. (4.28) with (4.25), one can interpret the angles α and β as *the values of eccentric anomaly on the rectilinear ellipse* between P_1 and P_2 *having the same value of* a, c, *and* $r_1 + r_2$.

The geometric interpretation of α and β and the quadrants for these angles then follows the classical geometric interpretation of eccentric anomaly encountered previously (Sec. 2.2). As shown in Fig. 4.6, one constructs the auxiliary circle of radius a centered at the center R of the rectilinear ellipse, located a distance $d = s - a$ from P_2. Points Q_1 and Q_2 are the intersections of lines normal to the chord through P_1 and P_2 with the auxiliary circle. The principal value angles α_0 and β_0 are shown in Fig. 4.6. Also shown is the fact that the *difference* $\alpha - \beta$ (regardless of quadrant) is equal to the difference in the values of eccentric anomaly on the *original* elliptic path between points P_1 and P_2 [see Eq. (4.21)].

Figure 4.7 shows separately the four elliptic arcs originally shown in Fig. 4.4, along with the corresponding geometric interpretations of the angles α and β. The top two figures depict the case $\theta \le \pi$ and are characterized by $\beta = \beta_0 \le \pi$. The top figure corresponds to the shorter transfer time for the given transfer angle for which $\alpha = \alpha_0 \le \pi$. The second figure corresponds to the longer transfer time, for which $\alpha = 2\pi - \alpha_0 \ge \pi$.

For a given transfer angle the behavior of the transfer time $t_2 - t_1$ as a function of semimajor axis is shown in Fig. 4.8. For $a > a_m$ the two elliptic arcs are characterized by one having a transfer time *defined as* t_F, which is less than t_m, the time on the minimum energy arc. The other has a transfer time *defined as* $t_F^\#$, which is greater than t_m. The value of t_m is easily

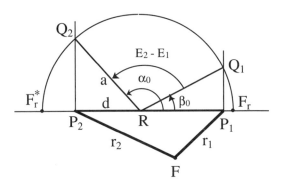

Fig. 4.6 Interpretations of Angles α_0 and β_0

determined from Eq. (4.25) using the fact that $a_m = s/2$ [Eq. (4.4)]. Equations (4.26) and (4.27) then tell us that

$$\alpha_m = \pi, \quad \sin(\beta_m/2) = \left[\frac{s-c}{s}\right]^{1/2} \tag{4.29}$$

and the time is given by

$$\sqrt{\mu} t_m = \left[\frac{s^3}{8}\right]^{1/2} (\pi - \beta_m + \sin \beta_m) \tag{4.30}$$

where for $0 \le \theta \le \pi$, $\beta_m = \beta_{m_0}$, and for $\pi \le \theta \le 2\pi$, $\beta_m = -\beta_{m_0}$. The limiting value of transfer time along the lower part of the curve of Fig. 4.8 is the *parabolic transfer time* t_p, which is approached asymptotically as $a \to \infty$. This will be discussed in more detail shortly.

Returning to Fig. 4.7, the lower two figures correspond to $\theta \ge \pi$ for which $\beta = -\beta_0$ with $\alpha = \alpha_0$ on the shorter transfer time arc and $\alpha = 2\pi - \alpha_0$ on the longer time arc.

To summarize, a single equation (4.25) can be used for all elliptic arcs with $0 \le \theta < 2\pi$, with the values of α and β determined from the principal values $0 \le \beta_0 \le \alpha_0 \le \pi$ as follows:

$$0 \le \theta < \pi, \quad \beta = \beta_0 \tag{4.31a}$$

$$\pi \le \theta < 2\pi, \quad \beta = -\beta_0 \tag{4.31b}$$

$$t_2 - t_1 = t_F \le t_m, \quad \alpha = \alpha_0 \tag{4.32a}$$

$$t_2 - t_1 = t_F^\# > t_m, \quad \alpha = 2\pi - \alpha_0 \tag{4.32b}$$

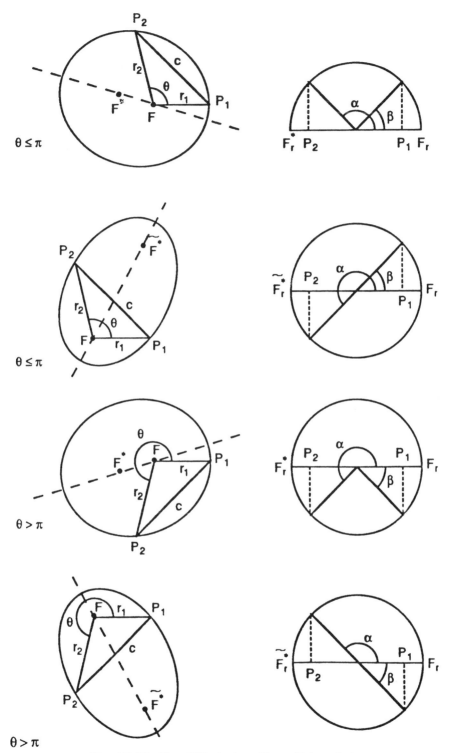

Fig. 4.7 The Four Elliptic Arcs (Same Value of a)

Note that for the special case $\theta = \pi$, $s = c$ and $\beta_0 = 0$; therefore, β is continuous at $\theta = \pi$. Also, for $t_2 - t_1 = t_m$, as mentioned in Eq. (4.29) $\alpha_0 = \pi$, and α is continuous at the transfer time t_m.

The equation describing the parabolic transfer time between specified endpoints is called *Euler's equation* (not that Euler needs another equation named after him) because this special case was published by Euler in 1743, almost 20 years before Lambert's more general result. It can be obtained by carefully taking the limit of the Lambert equation for an elliptic orbit as $a \to \infty$ (Prob. 4.3). The result can be compactly written using the *signum function*, sgn, defined by

$$\text{sgn}(x) = \begin{cases} 1 \text{ for } x > 0 \\ -1 \text{ for } x < 0 \end{cases} \quad (4.33)$$

Euler's equation is then

$$\sqrt{\mu}\,(t_2 - t_1) = \sqrt{\mu}\,t_p = \frac{\sqrt{2}}{3}\,[s^{3/2} - \text{sgn}(\sin\theta)(s-c)^{3/2}] \quad (4.34)$$

where the term sgn (sinθ) automatically accounts for the required sign change going from transfer angles less than π to greater than π. This equation yields a value of $t_p = 72$ days for the earth–Mars transfer geometry in Fig. 4.8. The value of t_p is important, since for an elliptic transfer to exist between specified endpoints, the transfer time must be greater than t_p.

A universal formulation of Lambert's equation exists using the special functions S and C described in Chap. 2. The Battin-Vaughan algorithm [4.4] and the Gooding procedure [4.5] are excellent methods for the iterative solution to Lambert's equation for the transfer orbit given the transfer time, is based on universal variables. This application is discussed in more detail in Sec. 4.6. A complete analysis of Lambert's problem in universal variables is given in [4.3].

Example 4.1

Consider the example transfer depicted in Fig. 4.8 with $r_1 = 1$ au, $r_2 = 1.524$ au, and $\theta = 75°$. The terminal radii represent an earth–Mars transfer orbit. For this geometry the chord $c = 1.592$ au, and the semiperimeter $s = 2.058$ au. From Eq. (4.4) the value of a_m is determined to be 1.03 au, as shown in Fig. 4.8. From Eqs. (4.29) and (4.30) this yields a value of time on the minimum energy ellipse of $t_m = 3.117$ canonical time units ($\mu = 1$) that corresponds to 181 days.

For a transfer time $t_2 - t_1$ equal to 115 days = 1.978 time units, Eq. (4.25) can be solved (by numerical iteration) for the value of a. Because the value of θ is less than 180°, β is simply β_0 as indicated in Eq. (4.31a). Also,

Lambert's Problem

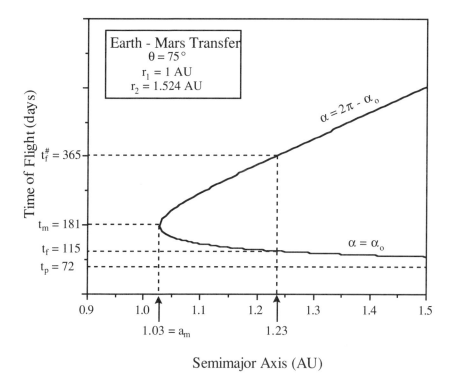

Fig. 4.8 Transfer Time vs. Semimajor Axis

because the specified transfer time is less than t_m, the solution for a lies on the lower portion of the curve in Fig. 4.8, and the specified time is denoted by t_F. The value of α that applies in this case is simply α_0 as indicated in Eq. (4.32a). The value of a obtained is 1.232 au, which corresponds to $\alpha_0 = 2.305 = 132.1°$ and $\beta_0 = 0.900 = 51.6°$. Thus the unique value of a has been determined for the specified transfer time.

The other value of transfer time for this same value of a lies on the upper portion of the curve in Fig. 4.8, corresponding to the transfer time $t_F^\#$. To determine this value, $\alpha = 2\pi - \alpha_0$ in Eq. (4.25), as indicated in Eq. (4.32b). The value obtained is $t_F^\# = 6.279$ time units = 365 days.

4.5 The Terminal Velocity Vectors

The terminal velocity vectors $\mathbf{v}_1$ at $\mathbf{r}_1$ and $\mathbf{v}_2$ at $\mathbf{r}_2$ can be conveniently expressed in terms of a set of skewed unit vectors, one along the *local* radius vector and the other along the chord. Specifically, let

$$\mathbf{u}_1 \equiv \frac{\mathbf{r}_1}{r_1}$$

$$\mathbf{u}_2 \equiv \frac{\mathbf{r}_2}{r_2} \tag{4.35}$$

$$\mathbf{u}_c \equiv \frac{(\mathbf{r}_2 - \mathbf{r}_1)}{c}$$

as shown in Fig. 4.9.

It can be shown that the velocity vector $\mathbf{v}_1$ can be expressed as

$$\mathbf{v}_1 = (B + A)\mathbf{u}_c + (B - A)\mathbf{u}_1 \tag{4.36}$$

where

$$A = \left[\frac{\mu}{4a}\right]^{1/2} \cot\left(\frac{\alpha}{2}\right) \tag{4.37a}$$

$$B = \left[\frac{\mu}{4a}\right]^{1/2} \cot\left(\frac{\beta}{2}\right) \tag{4.37b}$$

with the values of α and β being determined from Eqs. (4.26) and (4.27) with quadrant modifications given by Eqs. (4.31) and (4.32).

Equation (4.36) can be used to determine the terminal velocity vectors $\mathbf{v}_1$ and $\mathbf{v}_2$ for *all* the transfer ellipses having a given value of semimajor axis a. For specified values of the transfer angle $\theta \leq \pi$ and semimajor axis a, the principal values α_0 and β_0 in Eqs. (4.37) yield the correct components of $\mathbf{v}_1$ for the case $t_2 - t_1 = t_F < t_m$. The initial velocity $\mathbf{v}_1^\#$ on the other transfer ellipse having the same value of a, but with $t_2 - t_1 = t_F^\# > t_m$ is obtained by using $\alpha = 2\pi - \alpha_0$ in Eq. (4.37). This has the effect of merely changing the algebraic sign of the coefficient A in Eq. (4.36). This change in sign is equivalent to *interchanging the chordal and radial components* of the velocity vector in Eq. (4.36). Thus, the components of $\mathbf{v}_1^\#$ are easily obtained from the components of $\mathbf{v}_1$, and, because the value for a is the same for both, $|\mathbf{v}_1| = |\mathbf{v}_1^\#|$ (see Fig. 4.4).

The components of the final velocity $\mathbf{v}_2$ at $\mathbf{r}_2$ can also be obtained using Eq. (4.36) by considering a transfer backward in time from P_2 to P_1. In this context the velocity vector $-\mathbf{v}_2$ is the "initial" velocity, the chordal unit vector toward the final point is $-\mathbf{u}_c$, and the "initial" radial unit vector is $\mathbf{u}_2$. Equation (4.36) then becomes:

Lambert's Problem

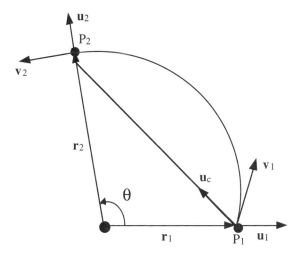

Fig. 4.9 Unit Vector Definitions

$$-\mathbf{v}_2 = (B + A)(-\mathbf{u}_c) + (B - A)\mathbf{u}_2 \qquad (4.38)$$

or,

$$\mathbf{v}_2 = (B + A)\mathbf{u}_c - (B - A)\mathbf{u}_2 \qquad (4.39)$$

Comparing Eqs. (4.36) and (4.39), one can see that *the chordal components of the terminal velocities $\mathbf{v}_1$ and $\mathbf{v}_2$ are equal, whereas the local radial components are the negatives of each other.*

Finally, the terminal velocities on the transfer ellipse for which $\pi < \theta < 2\pi$ are obtained by changing the sign on β, as in Eq. (4.31). This has the effect of changing the sign of the coefficient B in Eq. (4.36), which is equivalent to replacing the chordal component by the negative of the local radial components, and vice-versa. This makes sense, because the velocity vectors for $\theta > \pi$ and transfer time t_F are simply the negatives of the velocity vectors for $\theta < \pi$ and transfer time $t_F^\#$ (see Fig. 4.4).

In the typical application, for which the transfer angle θ and the transfer time $t_2 - t_1$ are specified, the transfer ellipse and the corresponding terminal velocity vectors are unique. The appropriate values of α and β must be used to obtain the correct orbit and the terminal velocities.

It should be noted that for $\theta = \pi$ Eq. (4.36) is indeterminant because the chordal unit vector and the local radial unit vectors become parallel. In this case, an alternate form of the velocity vector equation must be used.

Example 4.2

The velocity vectors corresponding to the transfer shown in Fig. 4.8 utilize the unit vectors: $\mathbf{u}_1^T = [1\ 0\ 0]$, $\mathbf{u}_2^T = [0.2588\ 0.9659\ 0]$, and $\mathbf{u}_c^T = [-0.3804\ 0.9248\ 0]$, which yields, using Eqs. (4.36) and (4.39):

$$\mathbf{v}_1^T = [0.3015\ 1.0476\ 0]\ ;\ \mathbf{v}_2^T = [-0.6205\ 0.3401\ 0]$$

While not a direct result from the Lambert problem, it can be noted that the universal variable x at point P_1 is, of course, 0, and the value at point P_2 is determined from the universal Kepler's Eq. (2.39) to be $x = 1.5600$.

4.6 Applications of Lambert's Equation

The typical application of Lambert's Eq. (4.25) is to determine the orbit and the terminal velocity vectors for specified $\mathbf{r}_1$, $\mathbf{r}_2$, and $t_2 - t_1$. The procedure is as follows:

1. Calculate the parabolic transfer time t_p using Eq. (4.34). If the specified $t_2 - t_1 > t_p$, an elliptic transfer orbit exists. Otherwise the orbit must be parabolic or hyperbolic.

2. Calculate t_m using Eq. (4.30) and note whether $t_2 - t_1$ is greater than or less than t_m. This determines the correct value of α in Eq. (4.32). The correct value of β is determined by the value of the transfer angle θ through Eq. (4.31).

3. Iteratively solve Lambert's Eq. (4.25) for the unique value of semimajor axis a. Standard iteration algorithms can be used, but the powerful universal algorithms developed by Battin and Vaughan [4.4] and Gooding [4.5] are recommended.

4. Once the value of a is determined, the terminal velocity vectors can be determined using Eq. (4.36) as discussed in Sec. 4.5.

The value of the eccentricity of the transfer orbit is best determined by first obtaining the parameter:

$$p = \frac{4a\,(s - r_1)\,(s - r_2)}{c^2} \sin^2\left[\frac{\alpha + \beta}{2}\right] \tag{4.40}$$

where α and β are determined as before by Eqs. (4.31) and (4.32). Then e can be determined from a and p using $p = a(1 - e^2)$.

One further aspect of the application is the possibility of long-duration arcs for which the transfer time is long enough so that multiple revolution solutions exist, for which $\theta \geq 2\pi$. In this case the transfer orbit is nonunique. One can always find a transfer of sufficiently large a that $\theta < 2\pi$ for the given transfer time, but there are a total of $2N + 1$ distinct solutions if the transfer time is long enough to allow N complete revolutions of the focus, where $2N\pi \leq \theta < 2(N+1)\pi$. The transfer time for $N \geq 1$ is related to the transfer time for $N = 0$ [Eq. (4.25)] by simply adding the term NT, where T is the period of the transfer orbit [Eq. (1.41)]:

$$\sqrt{\mu}\,(t_2 - t_1) = a^{3/2}\,[\,2N\pi + \alpha - \beta - (\sin\alpha - \sin\beta)\,] \qquad (4.41)$$

References

4.1 Battin, R. H., Fill, T. J., and Shepperd, S. W., "A New Transformation Invariant in the Orbital Boundary-Value Problem," *Journal of Guidance and Control* **1**, 1, Jan-Feb. 1978, pp. 50–55.

4.2 Prussing, J. E., "A Geometrical Interpretation of the Angles Alpha and Beta in Lambert's Problem," *Journal of Guidance and Control*, **2**, 5, Sept-Oct. 1979, pp. 442–443.

4.3 Battin, R. H., *An Introduction to the Mathematics and Methods of Astrodynamics*, American Institute of Aeronautics and Astronautics, New York, 1987.

4.4 Battin R. H., and Vaughan, R. M., "An Elegant Lambert Algorithm," *Journal of Guidance, Control and Dynamics*, **7**, 6, Nov–Dec. 1984, pp. 662–670.

4.5 Gooding, R. H., "A Procedure for the Solution of Lambert's Orbital Value Problem," *Celestial Mechanics*, **48**, 1990, pp. 145–165.

Problems

4.1 a) For a given space triangle, determine expressions for the terminal velocity vectors $\mathbf{v}_{1_m}$ and $\mathbf{v}_{2_m}$ on the minimum energy orbit between P_1 and P_2 in terms of the unit vectors $\mathbf{u}_c$, $\mathbf{u}_1$, and $\mathbf{u}_2$.
b) Interpret the directions of these velocity vectors geometrically in terms of the unit vector directions.

4.2 Consider the earth and Jupiter to be in coplanar circular orbits of radii 1 au and 5.2 au, respectively.
a) Considering the transfer angle θ as a variable, determine the range

of values of a_m for all the possible earth–Jupiter transfer ellipses.
b) For $\theta = 150°$ and $a = 5$ au, calculate the values of a_m (in au), t_m, t_F, $t_F^\#$, and t_p (in years).
c) Calculate $\mathbf{v}_1$ and $\mathbf{v}_1^\#$ (in EMOS) for the two transfer ellipses of (b),
d) Calculate the magnitudes of $\mathbf{v}_1$ and $\mathbf{v}_1^\#$,
e) calculate p and $\tilde{p}$ (in au) along with e and $\tilde{e}$.
f) for the two ellipses, perform the graphical construction for α and β described in the text.

4.3* Determine the expression (4.34) for t_p, the transfer time on a parabolic orbit between points P_1 and P_2. Start with Eq. (4.25) for an elliptic orbit and proceed to the limit as $a \to \infty$. Be sure to account for the two cases $\theta \le \pi$ and $\theta > \pi$.

4.4 Calculate the sum of t_F for $\theta < \pi$ and $t_F^\#$ for $\theta > \pi$ for a given elliptic orbit and interpret your result using Fig. 4.4.

4.5 Show that $p_m = 2(s-r_1)(s-r_2)/c = r_1 r_2 (1-\cos\theta)/c$.

4.6 Specialize the expressions for α, β, and $t_2 - t_1$ to the case of a *circular* arc of radius r_c and transfer angle θ.

4.7 For the case $r_1 = r_2 \equiv r_o$ and an arbitrary transfer angle θ,
a) Construct the locus of the focus.
b) For a value of a equal to r_o determine the values of e and $\tilde{e}$ and the corresponding values of p and $\tilde{p}$.

4.8 Determine the terminal velocity vectors $\mathbf{v}_1$ and $\mathbf{v}_2$ [Eqs. (4.36) and (4.39)] for a parabolic orbit. Accomplish this by evaluating the variables A and B in Eqs. (4.37a) and (4.37b) in the limit as $a \to \infty$.

5
Rocket Dynamics

5.1 Introduction

Beside gravity, the other major force that acts on a spacecraft is rocket thrust. There are a variety of types of rocket engines, typically categorized as either high- or low-thrust engines based on the magnitude of the thrust acceleration compared to the local gravitational acceleration. High-thrust engines can provide thrust acceleration magnitudes significantly higher than the local gravitational acceleration, while the value is typically several orders of magnitude less than gravity for a low-thrust engine. With either type of engine, the rocket mass is not constant, but decreases due to the fact that some mass is expelled out of the rocket nozzle to provide thrust, which is the reaction force in the opposite direction.

5.2 The Rocket Equation

In order to obtain the equation of motion for a system having a time-varying mass, one utilizes the fundamental form of Newton's Second Law, namely that the time rate of change of linear momentum relative to an inertial frame of reference is equal to the net external force acting on the system.

For the case of the rocket shown in Fig. 5.1 the variable m is the instantaneous mass of the rocket and $\mathbf{v}$ is the vector velocity of the center of mass of the rocket relative to an inertially fixed point labeled O.

The vector $\mathbf{c}$ is the effective exhaust velocity of the engine. The exhaust velocity is the velocity of the expelled particles *relative to the rocket*; the adjective *effective* implies that the effect of any nonzero exit pressure at the nozzle has been compensated for, in the case of a chemical rocket engine. This will be explained later.

For convenience, let us define the mass flow rate b as

$$b = -\dot{m} \geq 0 \qquad (5.1)$$

where $\dot{m}$ is negative because the mass m of the rocket is decreasing with time as the engine is operating.

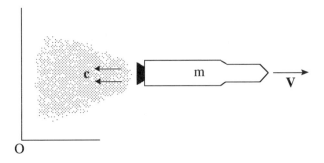

Fig. 5.1 Rocket Dynamic Variables

One next evaluates the total linear momentum of the system at some time t and compares it with the value at a slightly later time $t + \Delta t$. The change in momentum over the time interval Δt is then related to the net external force acting on the system.

At time t the linear momentum of the system is simply $m\mathbf{v}$. At time $t + \Delta t$ the linear momentum is the sum of two parts: the momentum of the rocket of slightly different mass and velocity, and the momentum of the particles exhausted during the time interval Δt.

$$(m - b\,\Delta t)(\mathbf{v} + \Delta\mathbf{v}) + (b\,\Delta t)(\mathbf{v} + \mathbf{c}) \tag{5.2}$$

In Eq. (5.2) the quantity $b\,\Delta t$ is the amount of mass lost by the rocket, and is exactly equal to the mass of the exhausted particles. $\Delta\mathbf{v}$ is the change in velocity of the rocket and $\mathbf{v} + \mathbf{c}$ is the vector velocity of the exhausted particles *relative to the inertial reference point O.*

By Newton's Second Law the change in linear momentum is equal to the applied mechanical impulse:

$$(m - b\,\Delta t)(\mathbf{v} + \Delta\mathbf{v}) + (b\,\Delta t)(\mathbf{v} + \mathbf{c}) - m\mathbf{v} = \mathbf{F}_{ext}\,\Delta t \tag{5.3}$$

Expanding and simplifying:

$$m\,\Delta\mathbf{v} - b\,\Delta t\,(\Delta\mathbf{v} - \mathbf{c}) = \mathbf{F}_{ext}\,\Delta t \tag{5.4}$$

Dividing by Δt and proceeding to the limit as $\Delta t \to 0$:

$$m\dot{\mathbf{v}} = -b\mathbf{c} + \mathbf{F}_{ext} \tag{5.5}$$

Rocket Dynamics

Equation (5.5) is called the *rocket equation* and the term $-b\mathbf{c}$ is the force called the *thrust* of the rocket, directed opposite to the exhaust velocity.

The first term on the right-hand side of Eq. (5.5) is the thrust of the rocket, which depends on both the mass flow rate and the exhaust velocity. The second term is the total *external* force due to all effects other than thrust, such as gravity or atmospheric drag. Note that in Eq. (5.3) the thrust appears as an *internal* force because our definition of the system boundary includes the exhausted particles. In any event, Eq. (5.5) describes the acceleration $\dot{\mathbf{v}}$ of the rocket under the influence of thrust and other forces.

The relationship between the *effective* exhaust velocity c and the *actual* exhaust velocity c_a for a chemical rocket can now be explained. The thrust magnitude bc is equal to $bc_a + (p_e - p_\infty)A_e$, where p_e is the pressure of the exhaust at the nozzle exit, p_∞ is the outside ambient pressure (which has the value 0 in a vacuum), and A_e is the nozzle exit area. If the exhaust gases are not fully expanded at the exit, then there is an additional contribution to the thrust besides bc_a, which is accounted for by defining the effective exhaust velocity c.

5.3 Solution of the Rocket Equation in Field-Free Space

First, let us consider the solution of Eq. (5.5) in the simple case $\mathbf{F}_{ext} = 0$ (i.e., field-free space). In this case the only force acting is the rocket thrust. As we will see, this is a useful approximation for high-thrust engines. Recalling that $b = -\dot{m}$ [Eq. (5.1)] one can write Eq. (5.5) in differential form as

$$m\, d\mathbf{v} = \mathbf{c}\, dm \tag{5.6}$$

or

$$d\mathbf{v} = \mathbf{c}\, \frac{dm}{m} \tag{5.7}$$

Assuming a constant exhaust velocity $\mathbf{c}$, Eq. (5.7) can be integrated to yield the velocity change $\Delta \mathbf{v}$.

The assumption of a constant magnitude for the exhaust velocity c is valid for a large class of engines, including *chemical, high-thrust* devices. In this case the value of c depends on pressure and temperature differences between the combustion chamber and the nozzle exit. Constant exhaust velocity low-thrust engines also exist which are so-called thrust-limited devices, such as a nuclear engine. We also assume for simplicity that the thrust direction is constant. Integrating Eq. (5.7) yields

$$\Delta \mathbf{v} = \mathbf{v} - \mathbf{v}_o = -\mathbf{c} \ln\left[\frac{m_o}{m}\right] \tag{5.8}$$

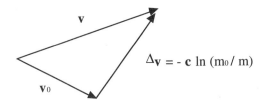

Fig. 5.2 Velocity Change Due to Thrust

where the term m_o/m is called the *mass ratio*. This velocity change is shown in Fig. 5.2 in the direction opposite to **c**.

Note that in the field-free case the velocity change $\Delta \mathbf{v}$ is independent of the time interval over which the thrust occurs. The fact that the magnitude Δv depends only on the amount of propellant consumed, $\Delta m = m_o - m$, and the exhaust velocity can be seen by rewriting Eq. (5.8) in the form

$$\frac{\Delta m}{m_o} = 1 - e^{-\Delta v/c} \tag{5.9}$$

Figure 5.3 depicts the relationship given in Eq. (5.9)

Note that the function shown in Fig. 5.3 is monotonic. This has important ramifications in determining mission propellant requirements because one can simply use the size of a required velocity change as a measure of the amount of fuel required. This is especially convenient because in orbit mechanics analysis mass is not explicitly considered, because dynamic quantities are most conveniently expressed per unit mass. Velocity, on the other hand, is easily calculated using the formulas of orbital analysis, such as the vis–viva equation and the velocity expressions from Lambert's problem.

The quantity Δv is also referred to as the *characteristic velocity* or simply the "delta-vee" of the maneuver. The fact that total propellant requirements for several thrust maneuvers can be accounted for by simply adding the characteristic velocities for each maneuver can be seen by the example in Fig. 5.4. Due to the first thrust

$$\Delta v_1 = c \ln \left[\frac{m_o}{m_1} \right] \tag{5.10}$$

Because m_1 is both the final mass for the first burn and the initial mass for the second burn, the second thrust yields:

Rocket Dynamics

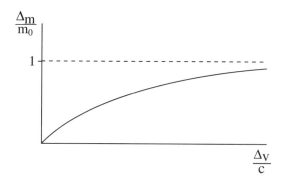

Fig. 5.3 Propellant Consumption vs. Velocity Change

$$\Delta v_2 = c \ \ln\left[\frac{m_1}{m_2}\right] \tag{5.11}$$

Thus

$$\Delta v_{total} = \Delta v_1 + \Delta v_2 = c \ \ln\left[\frac{m_o}{m_1}\right] + c \ \ln\left[\frac{m_1}{m_2}\right] = c \ \ln\left[\frac{m_o}{m_2}\right] \tag{5.12}$$

That is, the mass ratio due to both thrusts is accounted for by adding the two velocity change magnitudes. Clearly this result can be generalized to include any number of thrust periods.

In addition to exhaust velocity, a rocket engine is often characterized by its value of *specific impulse*, I_{sp}. This is equal to the total mechanical impulse imparted to the rocket divided by the weight (at the earth's surface) of propellant consumed. In an interval Δt:

$$I_{sp} = \frac{(cb)\,\Delta t}{(b\,\Delta t)\,g} = \frac{c}{g} \tag{5.13}$$

Thus I_{sp}, usually quoted in seconds, is merely a measure of the exhaust velocity c. In fact, in SI units the specific impulse is defined by dividing the impulse by the *mass* of propellant consumed, rather than the weight, and is identical to the exhaust velocity itself. However, specific impulse measured in seconds is in standard usage in the aerospace industry and further discussion of its fine points would be both pedantic and useless.

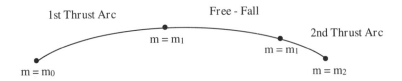

Fig. 5.4 A Two-Thrust Maneuver

As mentioned previously, rocket engines are typically classified as either high-thrust or low-thrust, depending on the magnitude of the thrust acceleration in comparison with 1 g. High thrust rockets are typically chemical rockets with a specific impulse in the 200–500 s range. The low end of this range corresponds to a World War II V-2 engine while the upper end is represented by the Advanced Space Engine at 476 s. From the definition of Eq. (5.13) it is evident that the corresponding range of exhaust velocities is roughly 2000–5000 m/s. A much newer high-thrust engine, the nuclear thermal rocket, has an I_{sp} of 850 s.

Low thrust engines, such as an ion engine, typically have very high specific impulses, of the order of 10^4 s because the ions are expelled at tremendously high speed (nearly 10^5 m/s). The question then arises, since thrust depends on exhaust velocity, how can an engine with such a high exhaust velocity be a low-thrust engine? The answer lies in the fact that the thrust is equal to the *product* of the *exhaust velocity* and the *mass flow rate*, bc [Eq. (5.5)]. The rate that mass is expelled out of the nozzle of an ion engine is extremely low because relatively few charged particles per second are expelled out the nozzle. By contrast, a high-thrust chemical engine, which has only a moderate exhaust velocity, has a high mass flow rate due to the large amount of combustion exhaust products flowing out the nozzle.

One final aspect of the solution to the rocket equation should be mentioned. While it is true that higher Δv magnitudes imply higher propellant consumption, the percentage changes in these variables can be very different. To see this, consider the solution to the rocket equation in the form

$$e^{-\Delta v/c} = \frac{m_f}{m_o} \qquad (5.14)$$

Rocket Dynamics

where m_f is the final mass. Examining first-order variations in $J = \Delta v$ and m:

$$-\frac{\delta J}{c} e^{-J/c} = \frac{\delta m_f}{m_o} \tag{5.15}$$

which can be expressed using (5.14) as

$$\frac{\delta m_f}{m_f} = -\frac{J}{c}\frac{\delta J}{J} \tag{5.16}$$

Thus the marginal change in final mass m_f is opposite in sign to the marginal change in $J = \Delta v$, but with proportionality factor $\Delta v/c$. The fact that the sign is opposite merely indicates the obvious fact that a decrease in Δv results in an increased final mass. However, the proportionality factor indicates that if $\Delta v > c$, then the marginal change in m_f is *greater* than the marginal change in Δv.

If the changes are large enough that a first-order, marginal analysis is not accurate, one must compare the nonlinear values directly. It can be shown that (see Prob. 5.8):

$$\frac{\Delta m_f}{m_f} = e^{\frac{-\Delta J}{c}} - 1 \tag{5.17}$$

As a numerical example, consider an interplanetary mission with a chemical rocket having an I_{sp} of 306 s ($c = 0.1$ EMOS). Two missions for solar system escape are compared in Prob. 7.3. The direct escape from Earth orbit requires a Δv magnitude of 0.294 EMOS, compared with a Jupiter gravity-assist, which has a Δv of 0.211 EMOS. Because the second Δv is 28 percent lower than the first, a first-order analysis is not accurate and one must use Eq. (5.17). The result is $\Delta J / c = -0.83$ and $\Delta m_f / m_f = 1.29$. Thus a modest 28 percent decrease in Δv results in a 129 percent increase in the final mass!

5.4 Solution of the Rocket Equation with External Forces

The solution to the rocket equation (5.5) in the presence of external forces is obtained similarly to (5.7) by writing

$$d\mathbf{v} = \mathbf{c}\frac{dm}{m} + \frac{\mathbf{F}_{ext}}{m} dt \tag{5.18}$$

Thus

$$\Delta \mathbf{v} = \mathbf{v} - \mathbf{v}_o = -\mathbf{c} \ln\left[\frac{m_o}{m}\right] + \int_{t_o}^{t} \frac{\mathbf{F}_{ext}}{m} dt \tag{5.19}$$

The velocity change depends, in addition to the mass ratio, on the mechanical impulse per unit mass provided by the external force. The magnitude of this term depends on the external force and on the length of the burn time. A simple case that can be calculated analytically is the vertical ascent of a rocket in a constant gravity field, for which

$$\dot{v} = \frac{cb}{m} - g \tag{5.20}$$

$$dv = -c\,\frac{dm}{m} - g\,dt \tag{5.21}$$

$$\Delta v = v - v_o = c\,\ln\left[\frac{m_o}{m}\right] - gt \tag{5.22}$$

The last term on the right-hand side of Eq. (5.22) is called a *gravity loss* term, because it represents a decrease in the velocity change due to the gravitational force, compared with the field-free space value.

As is evident from Eq. (5.22) the size of the gravity loss depends on the burn time. For a given thrust maneuver the burn time may be quite short for a very high-thrust engine, in which case the gravity loss is small and perhaps negligible compared with a lower-thrust engine that must burn for a longer time to perform the same velocity change. This is consistent with defining high thrust engines as having thrust accelerations of the same order of magnitude or higher than the local gravitational acceleration. Low-thrust engines have thrust accelerations several orders of magnitude smaller than the gravitational acceleration.

5.5 Rocket Payloads and Staging

The initial mass m_o of a rocket can be characterized by

$$m_o = m_p + m_s + m_L \tag{5.23}$$

where m_p is the *propellant mass*, m_s is the *structural mass*, and m_L the *payload mass*. The structural mass is defined as everything that is not propellant or payload mass. Assuming all the propellant is consumed, the mass at engine burnout is simply $m_s + m_L$.

The *mass ratio* at burnout can be expressed in terms of a variable Z defined as

$$Z = \frac{m_o}{m_s + m_L} \tag{5.24}$$

Rocket Dynamics

from which the characteristic velocity can be written from Eq. (5.8) as

$$\Delta v = c \ln Z \tag{5.25}$$

Also defined for convenience are two other nondimensional variables: the *payload ratio*

$$\lambda = \frac{m_L}{m_o - m_L} = \frac{m_L}{m_p + m_s} \tag{5.26}$$

and the *structural coefficient*

$$\varepsilon = \frac{m_s}{m_p + m_s} \tag{5.27}$$

The payload ratio is a measure of the payload mass in proportion to the mass of the rest of the rocket, and the structural coefficient characterizes the structural mass in proportion to the propellant mass, irrespective of payload. Various relationships can be derived involving these parameters, including the useful one

$$Z = \frac{1 + \lambda}{\varepsilon + \lambda} \tag{5.28}$$

Let us consider a numerical example of a typical single-stage chemical rocket. Assume the numerical values of the rocket parameters given in Ref. 5.1 of $I_{sp} = 311$ s. ($c = 3048$ m/s) and the masses in Table 5.1.

The mass of this vehicle is 80 percent propellant and its parameters are $\varepsilon = 1/7$, $\lambda = 1/14$, and $Z = 5$. Its characteristic velocity is

$$\Delta v = c \ln Z = 4906 \text{ m/s} \tag{5.29}$$

By comparison, circular orbit speed for LEO is approximately 7905 m/s, so even if there were no atmospheric or gravity losses, this would not be sufficient Δv to orbit a satellite. Let us consider the alternative of a two-stage version of this rocket with the assumption that the values of λ, ε, Z, and c are the same for the two stages; these are called *similar stages*. Assume also that $m_L = 1000$ units as before and that $m_o = 15000$ units, where now m_o is the sum of all the individual masses from both stages:

$$m_o = m_{s1} + m_{p1} + m_{s2} + m_{p2} + m_L \tag{5.30}$$

If the empty first stage is jettisoned prior to the second stage burn, the initial mass of the second stage is the "payload" of the first stage:

$$m_{o2} = m_{s2} + m_{p2} + m_L \tag{5.31}$$

Table 5.1 Rocket Parameters

m_L	1000	units
m_s	2000	units
m_p	12000	units
m_o	15000	units

Equal payload ratios requires

$$\lambda = \lambda_1 = \lambda_2 = \frac{m_{o2}}{m_o - m_{o2}} = \frac{m_L}{m_{o2} - m_L} \qquad (5.32)$$

or

$$m_{o2}^2 = m_o \, m_L \qquad (5.33)$$

Equal structural coefficients for the two stages requires

$$\varepsilon = \varepsilon_1 = \varepsilon_2 = \frac{m_{s1}}{m_o - m_{o2}} = \frac{m_{s2}}{m_{o2} - m_L} \qquad (5.34)$$

which yields the values, assuming $\varepsilon = 1/7$, shown in Table 5.2.

Note that the total structural mass is 2000 units and the total propellant mass is 12,000 units as in the single-stage rocket. Note also that the mass ratio of each stage is less than the single-stage rocket, but the total characteristic velocity is

$$\Delta v_T = \Delta v_1 + \Delta v_2 = 2c \ln Z = 6157 \text{ m/s} \qquad (5.35)$$

which is significantly larger (26 percent) than the single-stage value in Eq. (5.29), but still less than the 7905 m/s needed to establish a circular orbit.

The question arises: How can the two-stage Δv be larger when exactly the same amount of propellant, structure and payload are included? The answer, of course, lies in the fact that the empty first stage mass $m_{s1} = 1590$ units was jettisoned prior to the second-stage burn. This resulted in a lighter second stage being accelerated to a higher velocity using the available propellant.

Because the Δv for our two-stage example was not sufficient to establish a circular orbit, two options exist: (1) investigate the addition of more stages, or (2) decrease the payload mass. The first option can be eliminated

Table 5.2 Rocket Stage Masses

m_L	1000	units
m_{o2}	3873	units
m_{s1}	1590	units
m_{s2}	410	units
m_{p1}	9536	units
m_{p2}	2464	units
λ	0.348	
ε	0.1429	
Z	2.75	

for our example problem by utilizing the expression for the Δv for an n-stage rocket having similar stages in the limit as $n \to \infty$ (Prob. 5.7). This establishes an upper bound (actually a *supremum*) on the multistage Δv, which is

$$\text{limit } \Delta v \ (n \to \infty) = c \ (1 - \varepsilon) \ln \left[\frac{m_o}{m_L} \right] \tag{5.36}$$

which in our example case is 7075 m/s, still less that the required value of 7905 m/s.

To utilize the second option and accept a decreased payload mass, consider a three-stage rocket and require the Δv to be 7905 m/s. One can determine that the payload of the rocket can be only 532 units, rather than the 1000 units originally proposed. The details of the individual masses and rocket parameters for this case are given in Table 5.3. Note that to achieve the required Δv for the same structural mass of 2000 units, the propellant mass must be 83 percent of the total initial mass of the rocket. This increase in propellant mass is exactly equal to the decrease in payload mass.

In our simple example only 532 units or about 3.5 percent of the initial mass of the rocket is payload. This illustrates the difficulty in placing large masses in LEO and in delivering large payloads to the surface of the moon or to other planets.

In the example a structural coefficient $\varepsilon = 1/7 = 0.143$ was assumed for both stages. Recent advances in structural design methods and in the development of light-weight materials such as composites allow newer rockets to have more favorable values of ε. The newest commercial launch vehicle, Ariane IV (maiden flight June 15, 1988) has a first-stage $\varepsilon = 0.696$,

Table 5.3
One, Two, and Three-Stage Rockets
Similar Stages (Equal ε and λ)
$c = 3048$ m/s ($I_{sp} = 311$ s)

	1 stage	2 stage	3 stage	
			specified m_L	specified Δv
m_{o1}	15000	15000	15000	15000
m_{o2}	—	3873	6082	4926
m_{o3}	—	—	2466	1618
m_{s1}	2000	1589	1274	1393
m_{s2}	—	411	517	457
m_{s3}	—	—	209	150
m_{p1}	12000	9536	7644	8681
m_{p2}	—	2464	3099	2851
m_{p3}	—	—	1257	936
m_L	1000	1000	**1000**	532
ε	0.143	0.143	0.143	0.138
λ	0.0714	0.348	0.682	0.489
Δv (m/s)	4906	6157	6514	**7905**
$m_{s_{TOT}}$	2000	2000	2000	2000
$m_{p_{TOT}}$	12000	12000	12000	12468
Z	5	2.75	2.039	2.374

second-stage $\varepsilon = 0.0957$, and third-stage $\varepsilon = 0.1008$.

5.6 Optimal Staging

Suppose we wish to determine the optimal allocation of mass among the stages of a multistage rocket. The optimal allocation will minimize the propellant needed subject to a constraint that the total velocity change is equal to a specified value Δv_T. Minimizing a function subject to a constraint can be accomplished using a Lagrange multiplier. Before addressing the multistage rocket problem, let us consider a general mathematical example problem to introduce the concept of a Lagrange multiplier.

Rocket Dynamics

Suppose we want to minimize a function of two variables, $f(x,y)$. For a minimum of f, the function must be stationary with respect to variations in x and y:

$$df = \frac{\partial f}{\partial x} dx + \frac{\partial f}{\partial y} dy = 0 \tag{5.37}$$

Since dx and dy are arbitrary and independent, we must satisfy

$$\frac{\partial f}{\partial x} = \frac{\partial f}{\partial y} = 0 \tag{5.38}$$

Now suppose we want to minimize $f(x,y)$ subject to the constraint $g(x,y) = 0$. In addition to Eq. (5.37) we also have

$$dg = \frac{\partial g}{\partial x} dx + \frac{\partial g}{\partial y} dy = 0 \tag{5.39}$$

in order to hold the constraint. Thus dx and dy are not independent, but must satisfy Eq. (5.39), which along with Eq. (5.37) yields

$$\frac{f_x}{g_x} = \frac{f_y}{g_y} \tag{5.40}$$

where the subscript denotes a partial derivative.

If we call this ratio $-\eta$, then

$$f_x + \eta g_x = 0 \;,\;\; f_y + \eta g_y = 0 \tag{5.41}$$

Note that these equations are the same as would result if we minimized the function

$$h(x,y,\eta) = f(x,y) + \eta g(x,y) \tag{5.42}$$

yielding the three equations required for the solution of the three unknowns x, y, and (the Lagrange multiplier) η:

$$\frac{\partial h}{\partial x} = 0$$

$$\frac{\partial h}{\partial y} = 0 \tag{5.43}$$

$$\frac{\partial h}{\partial \eta} = g = 0$$

Returning to the multistage rocket problem, consider a two-stage rocket. In this case:

$$\Delta v_1 = c_1 \ln \frac{m_1 + m_2 + m_L}{\varepsilon_1 m_1 + m_2 + m_L} \tag{5.44a}$$

$$\Delta v_2 = c_2 \ln \frac{m_2 + m_L}{\varepsilon_2 m_2 + m_L} \tag{5.44b}$$

where the variable m_i represents the sum of the structural and propellant mass of stage i.

A practical problem is that given the payload mass m_L and the values of the c_i and ε_i, determine m_1 and m_2 to minimize the sum of the masses of the two stages $M = m_1 + m_2$ for a given $\Delta v_T = \Delta v_1 + \Delta v_2$.

The n-stage problem is to determine $m_1, m_2, \cdots m_n$ to minimize the function f:

$$f = M \equiv m_1 + m_2 + \cdots + m_n \tag{5.45}$$

subject to the constraint $g = 0$, given by:

$$\Delta v_1 + \Delta v_2 + \cdots + \Delta v_n - \Delta v_T = 0 \tag{5.46}$$

or,

$$\sum_{i=1}^{n} c_i \ln \frac{m_i + m_{i+1} + \cdots + m_n + m_L}{\varepsilon_i m_i + m_{i+1} + \cdots + m_n + m_L} - \Delta v_T = 0 \tag{5.47}$$

Unfortunately, conditions (5.43) applied to this problem:

$$\frac{\partial f}{\partial m_i} + \eta \frac{\partial g}{\partial m_i} = 0 \, , \, i = 1, 2, \ldots n \tag{5.48a}$$

$$g = 0 \tag{5.48b}$$

while yielding $n + 1$ independent equations for the $n + 1$ unknowns, namely the m_i and η, must be solved simultaneously and are quite complicated.

Note that in terms of the mass ratio Z_i, the constraint is very simple:

$$g = \sum_{i=1}^{n} c_i \ln Z_i - \Delta v_T = 0 \tag{5.49}$$

but unfortunately it is very difficult to express the function f in terms of the Z_i.

However, note that:

$$\frac{m_i + m_{i+1} + \cdots + m_n + m_L}{m_{i+1} + m_{i+2} + \cdots + m_n + m_L}$$

$$= \frac{(1 - \varepsilon_i)(m_i + \cdots + m_n + m_L)}{(1 - \varepsilon_i)(m_{i+1} + \cdots + m_n + m_L) + \varepsilon_i m_i - \varepsilon_i m_i}$$

Rocket Dynamics

$$= \frac{(1-\varepsilon_i)(m_i + m_{i+1} + \cdots + m_n + m_L)}{\varepsilon_i m_i + m_{i+1} + m_{i+2} + \cdots + m_n + m_L - \varepsilon_i(m_i + m_{i+1} + \cdots + m_n + m_L)}$$

$$= \frac{(1-\varepsilon_i)Z_i}{1-\varepsilon_i Z_i} \tag{5.50}$$

and note that

$$\frac{M+m_L}{m_2 + \cdots + m_n + m_L} \cdot \frac{m_2 + \cdots + m_n + m_L}{m_3 + \cdots + m_n + m_L} \cdots \frac{m_n + m_L}{m_L} = \frac{M+m_L}{m_L}$$

or,

$$\frac{M+m_L}{m_L} = \frac{(1-\varepsilon_1)Z_1}{1-\varepsilon_1 Z_1} \cdot \frac{(1-\varepsilon_2)Z_2}{1-\varepsilon_2 Z_2} \cdots \frac{(1-\varepsilon_n)Z_n}{1-\varepsilon_n Z_n} \tag{5.51}$$

Now, this function, $(M+m_L)/m_L$, is minimized when M is minimized, but is more simply expressed in terms of the Z_i.

Finally,

$$\ln \frac{M+m_L}{m_L} = \ln[(1-\varepsilon_1)Z_1] - \ln(1-\varepsilon_1 Z_1) + \cdots - \ln(1-\varepsilon_n Z_n)$$

$$= \ln(1-\varepsilon_1) + \ln Z_1 - \ln(1-\varepsilon_1 Z_1) + \cdots$$

$$= \sum_{i=1}^{n} [\ln(1-\varepsilon_i) + \ln Z_i - \ln(1-\varepsilon_i Z_i)] \tag{5.52}$$

and the function $\ln(M+m_L)/m_L$ is similarly minimized when $(M+m_L)/m_L$ and hence M are minimized.

Therefore, the problem can be restated as in [5.2]: Minimize

$$f = \sum_{i=1}^{n} [\ln(1-\varepsilon_i) + \ln Z_i - \ln(1-\varepsilon_i Z_i)] \tag{5.53}$$

subject to

$$g = \Delta v_T - \sum_{i=1}^{n} c_i \ln Z_i = 0 \tag{5.54}$$

For an extremum of f we require:

$$\frac{\partial f}{\partial Z_i} + \eta \frac{\partial g}{\partial Z_i} = [\frac{1}{Z_i} + \frac{\varepsilon_i}{1-\varepsilon_i Z_i}] + \eta[-\frac{c_i}{Z_i}] = 0, \quad i = 1, \cdots n \tag{5.55}$$

yielding

$$Z_i = \frac{\eta c_i - 1}{\eta c_i \varepsilon_i} \tag{5.56}$$

Therefore,

$$g = \Delta v_T - \sum_{i=1}^{n} c_i \ln\left[\frac{\eta c_i - 1}{\eta c_i \varepsilon_i}\right] = 0 \tag{5.57}$$

which may be solved, for example by trial and error, for η.

As an example, consider a three-stage rocket with the required $\Delta v_T = 10.42$ km/s. This velocity is geosynchronous transfer orbit (GTO) injection velocity (i.e., the speed required near the surface of the earth to place a satellite on a transfer orbit to geosynchronous radius). Suppose that the structural coefficients and exhaust velocities of the stages are given in Table 5.4 in which the c_i are given in units of kilometers per second.

Equation (5.57) is then solved by trial and error to yield $\eta = 0.334$, and Eq. (5.56) yields values for the mass ratios of $Z_1 = 2.385$, $Z_2 = 2.120$, and $Z_3 = 2.871$. The optimal mass for each stage can then be determined if the payload mass is given. For $m_L = 1000$ units,

$$Z_3 = \frac{m_3 + m_L}{\varepsilon_3 m_3 + m_l} \Longrightarrow m_3 = 2625 \tag{5.58}$$

$$Z_2 = \frac{m_2 + m_{3+} + m_L}{\varepsilon_2 m_2 + m_3 + m_L} \Longrightarrow m_2 = 5017 \tag{5.59}$$

$$Z_1 = \frac{m_1 + m_{2+} + m_3 + m_L}{\varepsilon_1 m_1 + m_2 + m_3 + m_L} \Longrightarrow m_1 = 14{,}791 \tag{5.60}$$

Then $m_o = m_1 + m_2 + m_3 + m_L = 23{,}433$ units is the minimum mass at launch of a vehicle that can give the payload mass of 1000 units the desired final velocity. Note that even with these "state-of-the-art" values of structural coefficients and exhaust velocities, the payload mass still represents only 4.3 percent of the mass at launch.

Table 5.4
Three-Stage Rocket Example

$\varepsilon_1 = 0.08$	$\varepsilon_2 = 0.09$	$\varepsilon_3 = 0.10$
$c_1 = 3.7$	$c_2 = 3.7$	$c_3 = 4.2$

Rocket Dynamics

References

5.1 Hill, P. G., and Peterson, C. R., *Mechanics and Thermodynamics of Propulsion*, Addison-Wesley Publishing Co., Reading, Mass., 1965, Sec. 10-4.

5.2 Peressini, A. L. "Lagrange Multipliers and the Design of Multistage Rockets," *UMAP Modules, Tools for Teaching*, Consortium for Mathematics and Its Applications, Inc., Arlington, Mass., 1986.

Problems

5.1 For the vertical ascent of a rocket in a constant gravity field of strength g, assume *constant* mass flow rate and exhaust velocity and determine an expression for the altitude $h(t)$.

5.2 At time $t = 0$, ignition of a rocket on a launch pad occurs. The thrust is equal to one-half the initial weight of the rocket (due to a design error) and the mass flow rate b is constant. Determine the time of lift-off and express your answer in terms of I_{sp}.

5.3 For the case of vertical ascent in a constant gravity field, obtain an expression for the burnout speed v_b and the burnout time t_b in terms of the mass ratio $Z = m_o/m_b$ and the parameter K defined by $bc = Km_o g$ ($K \geq 1$). Evaluate your expression for $Z = 5$, $K = 2$, $I_{sp} = 300$ s. Express your answers in terms of v_b/c and t_b/I_{sp}.

5.4 For the case treated in Prob. 5.3, obtain an expression for the *altitude* at burnout h_b and evaluate numerically.

5.5* For the case of Prob. 5.4, obtain an expression for the maximum altitude h_m in terms of h_b and v_b and plot the nondimensional variable h_m/cI_{sp} as a function of Z^{-1} for values of $K = 1$, 1.5, and 2.

5.6 Consider a rocket of three similar stages (equal λ, ε, c, and Z with $m_o = 15000$ units, $m_L = 1000$ units, $c = 3048$ m/s, and a total structural mass of 2000 units.
a) Calculate the values of m_{o2}, m_{o3}, and λ.
b) Calculate m_{s1}, m_{s2}, and m_{s3}.
c) Calculate m_{p1}, m_{p2}, and m_{p3}.
d) Determine the values of Z and the total Δv for the three-stage rocket.

5.7* Determine an expression for the total Δv for a rocket having n similar stages and determine the limiting value as $n \to \infty$ (Eq. 5.36).

5.8 Derive Eq. 5.17 of the text and verify that

$$\left| \frac{\Delta m_f}{m_f} \right| > \left| \frac{\Delta J}{J} \right|$$

if

$$\frac{J}{c} > \frac{\ln(1 + \Delta J/J)}{\Delta J/J}$$

5.9* For a rocket having three similar stages with the same total mass and total structural mass as Prob. 5.6, determine the payload mass for which the total velocity change is 7905 m/s and calculate the values of all masses and other important parameters which appear in Table 5.3.

5.10 a) For a rocket having n similar stages (i.e., having equal ε_i and equal c_i for all stages) Eq. (5.57) has a simple solution for the Lagrange multiplier η and hence for the mass ratio Z of each stage. Determine for $n = 2, 3, 4,$ and 5, the optimal value of m_o/m_L for the values $\Delta v_T = 10.15$ km/s, $\varepsilon_i = 0.10$ and $c_i = 2.866$ km/s.

b) For the given parameters and considering that each additional stage adds cost and complexity, what number of stages represents the best compromise between performance and complexity?

6
Impulsive Orbit Transfer

6.1 Introduction

The term *orbit transfer*, as used here, implies the general class of problems that includes interception and rendezvous as specific examples. This chapter continues the discussion in Chaps. 4 and 5. In Chap. 4 the velocity vector $\mathbf{v}_1$ was determined, which would take a vehicle from its initial point to a desired destination point. In Chap. 5 the mechanism for generating the required velocity changes using rocket thrust was discussed. Now in Chap. 6 the use of impulsive thrust to transfer between specific terminal orbits will be considered.

6.2 The Impulsive Thrust Approximation

The impulsive thrust approximation has application in the case of high-thrust rockets. To date, all the space missions flown, such as Apollo, Voyager, Space Shuttle, Galileo, and Magellan have employed high-thrust chemical engines. Low-thrust engines are currently in the development and testing phase.

The basic underlying idea behind the impulsive thrust approximation is that, due to the high thrust level, the burn times are very small compared with the time intervals between burns. The thrusts can then be idealized as having infinitely small duration, as shown in Fig. 6.1. Once the vehicle is in orbit, this is a very good approximation that is widely used. However, for the ascent phase from the surface of a planet, during which the vehicle is accelerated from rest to orbital speed, the losses due to the finite burn time must be included.

The strength of a thrust impulse is represented not by its amplitude, which is infinite, but by its time integral, which is geometrically represented by the area contained within the impulse. This area physically represents the magnitude of the instantaneous velocity change that occurs as a result of the thrust impulse. A simple way to see this is to define a *unit* impulse $\delta(t)$ (also called the *Dirac delta function*) as the limit of a finite pulse $p(t)$ of duration ε and amplitude $1/\varepsilon$, which is shown in Fig. 6.2. The pulse $p(t)$ has a unit time integral represented by the unit area contained within the pulse. The unit

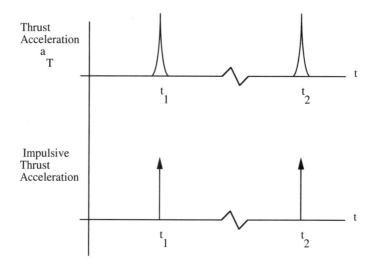

Fig. 6.1 Impulsive Approximation

impulse $\delta(t)$ is defined as

$$\delta(t) = p(t) \quad \text{in the limit as } \varepsilon \to 0 \tag{6.1}$$

Because the area enclosed by $p(t)$ is independent of ε, it remains unity in the limit.

A vector thrust acceleration impulse that provides an instantaneous vector velocity change $\Delta \mathbf{v}$ can be represented by $\Delta \mathbf{v}\,\delta(t)$. Because the change in vector velocity is finite, the position vector, being the time integral of the velocity vector, is continuous across a thrust impulse, as is shown in Fig. 6.3, for which the times t_k^- and t_k^+ represent times immediately before and after the thrust impulse at time t_k. A unit impulse that occurs at an arbitrary time t_k is represented by $\delta(t - t_k)$, that is, the impulse occurs at that time t for which the argument of the impulse function is equal to 0.

Because the thrust impulse is infinite in magnitude, the other, finite external forces acting have no effect and no gravity losses occur in the impulsive thrust approximation. The accuracy of this approximation has been studied by Robbins [6.1].

Impulsive Orbit Transfer

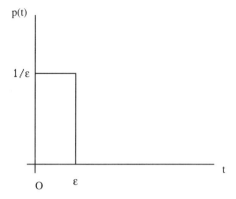

Fig. 6.2 Finite Pulse $p(t)$

An application of the impulsive thrust approximation is shown by Fig. 6.3. The velocity vector $\mathbf{v}^+ = \mathbf{v}(t_k^+)$ represents the desired velocity at the given point P; as an example, this could be the velocity vector $\mathbf{v}_1$ determined by solving Lambert's problem for the transfer orbit to a specified destination. The other velocity vector $\mathbf{v}^- = \mathbf{v}(t_k^-)$ shown is the vector velocity on the orbit that brought the vehicle to point P. The difference $\Delta \mathbf{v}$ determines the magnitude and direction of the required thrust impulse. As discussed in Chap. 5, the magnitude Δv is a direct measure of the propellant required to provide the velocity change.

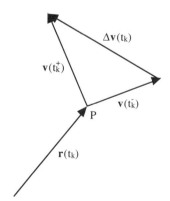

Fig. 6.3 Vector Velocity Change

Note that although the words sound similar, a minimum-fuel trajectory and a minimum-energy transfer orbit are not in general the same. The *energy* of the transfer orbit depends only on the values of **r** and **v**$^+$ on the transfer orbit immediately after the impulse. In contrast, the *fuel required* depends on the value of Δv and therefore *also* depends on the velocity **v**$^-$ *prior to* the impulse.

6.3 Two-Impulse Transfer between Circular Orbits

As an illustration of a class of orbit transfer problems, consider the simple case of circular terminal orbits of radii r_1 and r_2 as shown in Fig. 6.4.

The vehicle is initially in the circular orbit of radius r_1 and it is desired to place the vehicle in the circular orbit of radius r_2. The transfer orbit shown is an elliptic orbit that has the same focus as the terminal orbits. The orbit transfer shown can always be accomplished using two impulses as shown in Fig. 6.4. The first places the vehicle on the transfer orbit and the second circularizes the vehicle orbit at the final radius. Applications of this type of transfer are (1) *interplanetary transfer*, in which the planetary orbits, for example those of earth and Mars, are represented by circular orbits, and (2) *earth orbit transfer*, in which the circular terminal orbits could represent low-earth orbit (LEO) and geosynchronous orbit (GEO).

In order for a conic orbit such as that shown in Fig. 6.4 to be feasible as an impulsive transfer orbit, it must intersect both terminal orbits. This simple fact excludes all those conic orbits whose periapse lies outside the inner circular orbit and whose apoapse lies inside the outer circular orbit. Feasible conic transfer orbits satisfy the conditions:

$$r_p = \frac{p}{1+e} \leq r_1$$

$$r_a = \frac{p}{1-e} \geq r_2 \qquad (6.2)$$

where Eq. (1.36) has been used to express the periapse and apoapse radii in terms of the parameter p and eccentricity e of the transfer orbit. Equation (6.2) can be rewritten as:

$$p \leq r_1(1+e)$$

$$p \geq r_2(1-e) \qquad (6.3)$$

Each conic orbit about the given focus can be characterized by its values of p and e and can be represented as a point in the (p, e) plane. The inequalities given by Eq. (6.3) define a region of the plane that contains all

Impulsive Orbit Transfer 103

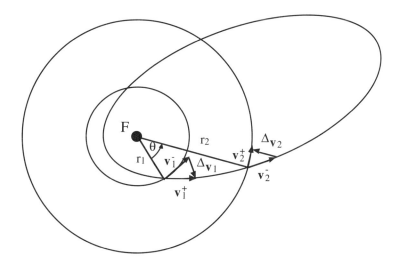

Fig. 6.4 Two-Impulse Transfer Orbit

the feasible transfer orbits between the given circular terminal orbits. As shown in Fig. 6.5 the boundaries of this region are straight lines, which is the reason for using the variable p rather than a. The region of feasible transfer orbits contains ellipses for $0 < e < 1$, parabolas for $e = 1$, and hyperbolas for $e > 1$, with *all* the rectilinear orbits located at point R for which $e = 1$ and $p = 0$. (However, not all rectilinear orbits are feasible; why not?).

Point H on Fig. 6.5 is interesting if only because it represents the least eccentric (most circular) of all the feasible transfer orbits. As will be seen in the next section, however, the transfer represented by point H is even more significant because it is the *Hohmann transfer*, which is the minimum-fuel solution for a large class of transfer problems.

6.4 The Hohmann Transfer

In 1925 Walter Hohmann published a monograph titled *Die Erreichbarkeit der Himmelskörper* [*The Attainability of Celestial Bodies*]. He conjectured that the minimum-fuel impulsive transfer orbit between coplanar circular orbits is the elliptic orbit that is tangent to both of the terminal orbits, as shown in Fig. 6.6.

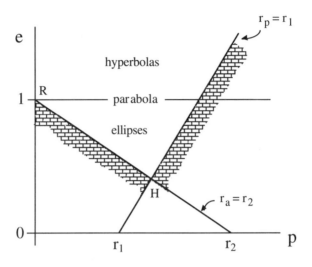

Fig. 6.5 Feasible Circle-to-Circle Transfer Orbits

The Hohmann transfer is an interesting counterexample of the legendary Murphy's Law because the semimajor axis, eccentricity, and transfer time are easier to calculate for this important transfer than for any other transfer orbit. The semimajor axis is determined by observing that in Fig. 6.6 $2a_H = r_1 + r_2$. Therefore,

$$a_H = \frac{(r_1 + r_2)}{2} \tag{6.4}$$

The semimajor axis is simply equal to the mean value of the terminal orbit radii, and Lambert's problem has been solved by inspection. The transfer time is simply half the period of the Hohmann ellipse:

$$t_H = \pi \left[\frac{a_H^3}{\mu} \right]^{1/2} = \pi \left[\frac{(r_1 + r_2)^3}{8\mu} \right]^{1/2} \tag{6.5}$$

and the eccentricity can easily be determined using the fact that the periapse radius is r_1 to be:

$$e_H = \frac{r_2 - r_1}{r_2 + r_1} \tag{6.6}$$

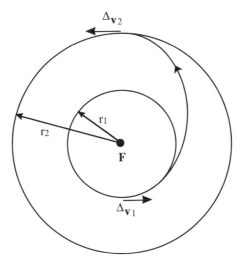

Fig. 6.6 The Hohmann Transfer

The basic properties of the Hohmann transfer can be summarized in nondimensional form in terms of $R \equiv r_2/r_1 \geq 1$ and the period T_1 of the inner circular terminal orbit as:

$$\frac{a_H}{r_1} = \frac{(1+R)}{2} \tag{6.7a}$$

$$\frac{t_H}{T_1} = \frac{\sqrt{2}}{8}(1+R)^{3/2} \tag{6.7b}$$

$$e_H = \frac{R-1}{R+1} \tag{6.7c}$$

The determination of the velocity changes Δv_1 and Δv_2 is left as an exercise for the reader (Prob. 6.1). One might think that the proof that the Hohmann transfer is the minimum-fuel transfer would also be easy, but it is not, primarily because this simple transfer must be compared to a general conic transfer orbit, for which the Δv calculations are much more complicated. A rigorous demonstration that the Hohmann transfer minimized fuel was not made until the 1960s. More recently, Palmore [6.2] published a brief proof showing the basic minimum-fuel property of the Hohmann transfer using ordinary calculus. In that proof use is made of the constant components of the velocity vector given in Eq. (3.13). In the 1960s Lawden [6.3] provided the theoretical basis for much of the subsequent minimum-fuel transfer research using the calculus of variations, research that continues to this day.

One can demonstrate the optimality of the Hohmann transfer among two-impulse circle-to-circle transfers rather simply as in [6.5] using the basic gradient idea of Palmore but with the variables e and p in Fig. 6.5 rather than the constant components of the velocity. The magnitude of a velocity change to depart or enter a circular orbit at $r = r_k$ where $k = 1,2$ is described by the law of cosines as:

$$(\Delta v)^2 = v^2 + v_c^2 - 2v_c v_\theta \quad (6.8)$$

where v_θ is the component of the velocity vector normal to the radius. Using Kepler's Second Law, Eq. (1.38), $v_\theta = h/r = (\mu p)^{1/2}/r$. In addition, from the vis–viva equation (1.50), $v_c^2 = \mu/r$ and, using $p = a(1 - e^2)$,

$$v^2 = \mu \left(\frac{2}{r} + \frac{e^2 - 1}{p} \right) \quad (6.9)$$

The total velocity change is then

$$\Delta v_T = \Delta v_1 + \Delta v_2$$

where Δv_1 occurs at $r = r_1$ and Δv_2 occurs at $r = r_2$. Thus Δv in Eq. (6.8) can be replaced by Δv_k corresponding to $r = r_k$. If one examines the gradient of the total velocity change with respect to the eccentricity of the transfer orbit, one obtains

$$\frac{\partial \Delta v_T}{\partial e} = \frac{\partial \Delta v_1}{\partial e} + \frac{\partial \Delta v_2}{\partial e} \quad (6.10)$$

and, from differentiating Eq. (6.8) using (6.9):

$$2\Delta v_k \frac{\partial \Delta v_k}{\partial e} = 2v_k \frac{\partial v_k}{\partial e} = \frac{2e\mu}{p}. \quad (6.11)$$

Thus,

$$\frac{\partial \Delta v_T}{\partial e} = \frac{e\mu}{p} (\Delta v_1^{-1} + \Delta v_2^{-1}) > 0 \quad (6.12)$$

Impulsive Orbit Transfer 107

The fact that the gradient of the total velocity change with respect to the eccentricity is *positive* means that in Fig. 6.5 at *any* point (p, e) in the interior of the feasible region the total velocity change can be *decreased* by *decreasing* the value of e while holding the value of p constant. This means that the minimum total velocity change solution will always lie on the cross-hatched boundary of the feasible region given by $r_a = r_2$ or $r_p = r_1$ or both.

What will be demonstrated is that, when the velocity change is restricted to the boundary, the gradient with respect to e remains positive. Thus the optimal solution lies at the minimum value of e on the boundary (i.e., at point H in Fig. 6.5), which represents the Hohmann transfer.

To do this, the expression for the total velocity change is restricted to the boundary by substituting for the value of p along each portion of the boundary, namely $p = r_2(1 - e)$ on the left part $(r_a = r_2)$ and $p = r_1(1 + e)$ on the right part $(r_p = r_1)$. Thus the total velocity change becomes a function of a *single* variable e. Letting $\Delta \hat{v}$ denote the velocity change restricted to the boundary, Eq. (6.8) becomes for the left part:

$$(\Delta \hat{v})^2 = \mu \left[\frac{2}{r} + \frac{e^2 - 1}{r_2(1-e)} \right] + \frac{\mu}{r} - 2\left(\frac{\mu}{r}\right)^{1/2} \frac{\left[\mu r_2(1-e)\right]^{1/2}}{r} \qquad (6.13)$$

Differentiating with respect to the single variable e and recalling that $R > 1$ and $0 < e < 1$:

$$2\Delta\hat{v}_1 \frac{d \Delta \hat{v}_1}{de} = \frac{\mu}{r_2}\left[R^{3/2} \frac{1}{(1-e)^{1/2}} - 1 \right] > 0 \qquad (6.14)$$

and

$$2\Delta\hat{v}_2 \frac{d \Delta \hat{v}_2}{de} = \frac{\mu}{r_2}\left[\frac{1}{(1-e)^{1/2}} - 1 \right] > 0 \qquad (6.15)$$

and thus, similar to Eq. (6.12) $d \Delta \hat{v}_T / de > 0$ on the left portion of the boundary.

Similarly, on the right portion of the boundary:

$$(\Delta \hat{v})^2 = \mu \left[\frac{2}{r} + \frac{e^2 - 1}{r_1(1+e)} \right] + \frac{\mu}{r} - 2\left(\frac{\mu}{r}\right)^{1/2} \frac{\left[\mu r_1(1+e)\right]^{1/2}}{r} \qquad (6.16)$$

Differentiating, one obtains:

$$2\Delta\hat{v}_1 \frac{d \Delta \hat{v}_1}{de} = \frac{\mu}{r_1}\left[1 - \frac{1}{(1+e)^{1/2}} \right] > 0 \qquad (6.17)$$

$$2\Delta\hat{v}_2 \frac{d\Delta\hat{v}_2}{de} = \frac{\mu}{r_1}\left[1 - \frac{1}{R^{3/2}(1+e)^{1/2}}\right] > 0 \qquad (6.18)$$

and thus, $d\Delta\hat{v}_T/de > 0$ on the right portion of the boundary also. It follows that the optimal solution lies at the point of minimum eccentricity on the boundary, namely at point H in Fig. 6.5, representing the Hohmann transfer.

What has been proved is that the Hohmann transfer is the absolute minimum-fuel two-impulse transfer between circular orbits.

6.5 Coplanar Extensions of the Hohmann Transfer

An interesting discussion of the Hohmann transfer, namely its extensions such as the *bi-parabolic* and *bi-elliptic* transfers and other minimum-fuel impulsive transfers is given by Edelbaum [6.4]. In 1959 three articles appeared independently by Edelbaum, Hoelker and Silber, and Shternfeld (references are given in Ref. 6.4), which showed that if the terminal radius ratio $R = r_2/r_1$ was sufficiently large, other transfers exist that require less fuel than the Hohmann transfer.

These other transfers are based on the extension of the Hohmann transfer called the *bi-elliptic* transfer, shown in Fig. 6.7.

The bi-elliptic transfer is a three-impulse transfer composed of two ellipses, separated by a midcourse tangential impulse located at a radius r_i as shown. A related transfer, the *bi-parabolic* transfer, is the limiting case as $r_i \to \infty$, for which both transfer segments are parabolic orbits separated by an impulse whose magnitude approaches 0 in the limit (Fig. 6.8).

The Δv magnitudes for the bi-parabolic transfer are easily calculated as

$$\Delta v_k = (\sqrt{2} - 1)\, v_{ck} \quad ; \quad k = 1, 2 \qquad (6.19)$$

where v_{ck} is the circular orbit speed in the terminal orbit of radius r_k. The total velocity change $\Delta v_{bp} = \Delta v_1 + \Delta v_2$ can be expressed in nondimensional form in terms of v_{c1} as:

$$\frac{\Delta v_{bp}}{v_{c1}} = (\sqrt{2} - 1)(1 + 1/\sqrt{R}) \qquad (6.20)$$

By comparing the fuel cost of the bi-parabolic transfer given in Eq. (6.20) with the Hohmann transfer cost, it can be shown (Prob. 6.7) that for values of R greater than approximately 11.94, the bi-parabolic transfer requires less fuel. This critical value of R is determined as a root of the cubic equation:

$$R^3 - (7 + 4\sqrt{2})R^2 + (3 + 4\sqrt{2})R - 1 = 0 \qquad (6.21)$$

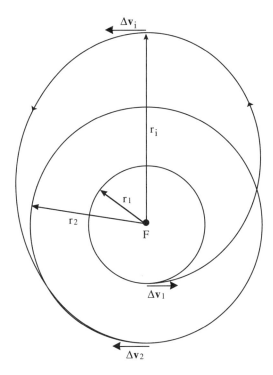

Fig. 6.7 The Bi-Elliptic Transfer

which has only one root larger than unity (recall our definition that $R \geq 1$).

Figure 6.9 displays the nondimensional fuel cost for the Hohmann and bi-parabolic transfers as a function of R, as well as two cases of the bi-elliptic transfer, for which the midcourse impulse is located at radii of $2r_2$ and $5r_2$.

An obvious disadvantage of the bi-parabolic transfer is that it requires infinite transfer time. This led researchers to investigate the finite time bi-elliptic transfer, which can also save fuel relative to the Hohmann transfer, but which has a transfer time that is greater than twice the Hohmann transfer time (see Fig. 6.7). The critical value of R that determines whether a bi-elliptic transfer requires less fuel than the Hohmann is the value at which the Hohmann Δv curve has its maximum value, as shown in Fig. 6.9. This value is a root of the cubic equation:

$$R^3 - 15R^2 - 9R - 1 = 0 \tag{6.22}$$

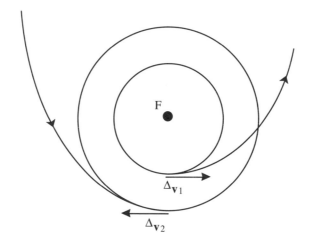

Fig. 6.8 The Bi-Parabolic Transfer

which has only one positive root, equal to approximately 15.58 (Prob. 6.8).

One other aspect of the Δv curves in Fig. 6.9 is worth mentioning: All the curves shown approach the nondimensional escape Δv, equal to $\sqrt{2} - 1$ in the limit as $R \to \infty$. The Hohmann Δv curve crosses this value at the value of R given by another cubic equation:

$$8R^3 - 33R^2 + 22R - 1 = 0 \tag{6.23}$$

which has only one root greater than unity, approximately $R = 3.3$ (Prob. 6.9). The significance of this number is that for any Hohmann transfer with $R > 3.3$, the total fuel cost for the two impulses is greater than the fuel cost to escape the center of attraction with a single impulse.

The minimum-fuel properties of the circle-to-circle coplanar transfer discussed earlier can now be summarized. For $1 < R < 11.94$ the absolute minimum-fuel transfer is the Hohmann transfer. (This range of values includes many practical applications such as LEO to GEO transfers and interplanetary transfers from earth to all the planets except the outermost three.) For $R > 11.94$, the bi-parabolic transfer is the absolute minimum-fuel transfer, although the Hohmann transfer remains as the minimum-fuel finite time two-impulse transfer. For $R > 15.58$ any bi-elliptic transfer for which

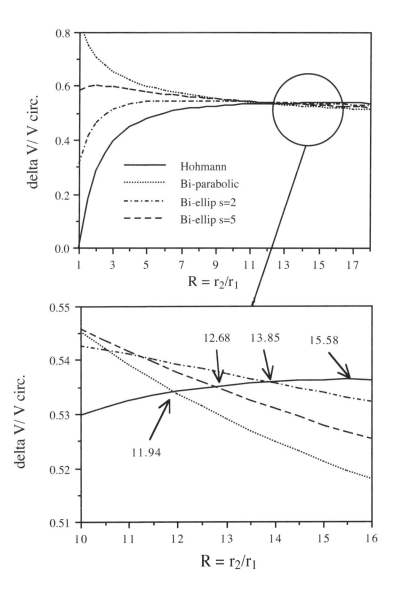

Fig. 6.9 Fuel Cost as a Function of R

the midcourse impulse lies outside the outer orbit ($r_i > r_2$) is more economical than the Hohmann transfer. For $11.94 < R < 15.58$ the bi-elliptic transfer is more economical than the Hohmann transfer only if the midcourse impulse location r_i is sufficiently large (Fig. 6.9).

The bi-elliptic transfer is significantly more advantageous in the noncoplanar case considered in the next section, but it is an interesting example of a minimum-fuel transfer that requires *more impulses* than are necessary to simply satisfy the boundary conditions of the problem.

6.6 Noncoplanar Extensions of the Hohmann Transfer

The noncoplanar, or inclined, orbit transfer problem provides additional aspects of minimum-fuel trajectories and also extends minimum-fuel results to a wide variety of practical applications that require a plane change in addition to changing the terminal radius.

A simple plane change can always be accomplished with a single impulse as shown in Fig. 6.10. In Fig. 6.10 a velocity change rotates the velocity vector through an angle θ, establishing a new orbit that is inclined to the original orbit, a process sometimes called *orbit cranking*. For the case in which $v^- = v^+ = v_c$ (a circular orbit), the magnitude of the velocity change is simply

$$\Delta v = 2 v_c \sin (\theta/2) \tag{6.24}$$

One might suspect that this maneuver is the obvious minimum-fuel "transfer" between inclined circular orbits of equal radius, but in fact, one can *always* save fuel over this one-impulse maneuver by using three impulses. This three-impulse transfer is a noncoplanar bi-elliptic transfer, discussed by Edelbaum [6.4] and shown in Fig. 6.11. For the optimum maneuver, small plane changes are made at the initial and final impulses with the majority of the total plane change made at the second impulse, located at a point outside the terminal orbit. The reason for making most of the desired

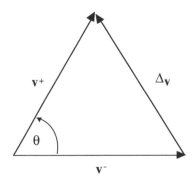

Fig. 6.10 A Single-Impulse Plane Change

plane change there is simply that the speed of the vehicle is less, and the Δv for a given plane change angle is proportional to the speed [Eq. (6.24)]. This is the reason that the bi-elliptic transfer is more useful in the inclined case than in the coplanar case. It is worth expending the Δv to travel to a larger radius because the saving in Δv to make the plane change at lower speed more than compensates for it, and the three impulses use less total fuel than the single impulse of Fig. 6.10. The reason that small, but nonzero plane changes are made at the terminal impulses is that when one is accelerating or decelerating the vehicle with an impulse, a differential or first-order small rotation of the velocity vector can be made with negligible cost.

A graph of the $\Delta v / v_c$ versus the plane change angle θ is shown in Fig. 6.12, with the optimum apoapse radius of the transfer ellipse designated. As shown, for values of θ less than 60.185°, the optimum transfer is bi-elliptic; for values of θ greater than 60.185° the optimum transfer is the bi-parabolic transfer, for which the cost of the plane change provided by the second impulse tends to 0 in the limit as $r_i \to \infty$.

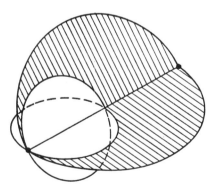

Fig. 6.11 Noncoplanar Bi-Elliptic Transfer between Circular Orbits (Copyright, 1966, American Institute of Aeronautics and Astronautics; reprinted with permission)

The total Δv cost of the inclined bi-parabolic transfer is simply:

$$\frac{\Delta v}{v_c} = 2(\sqrt{2} - 1) = 0.828 \qquad (6.25)$$

Note that from Eqs. (6.24) and (6.25) the bi-parabolic transfer has a smaller Δv than the single-impulse plane change for values of θ greater than 48.940° (see Fig. 6.12).

6.7 Conditions for Interception and Rendezvous

Let us again consider the simple case of coplanar circular terminal orbits. An interception (which is also required for a rendezvous) is more restrictive than a simple orbit transfer because the target must simultaneously be at the arrival point of the vehicle in the final orbit. This is equivalent to saying that at the time the vehicle departs the initial orbit, the target has to be in a specific location in the final orbit so that interception is achieved.

This requirement for interception determines the relative geometry between the vehicle in the initial orbit and the target in the final orbit at the instant of injection of the vehicle into the transfer orbit. The periods of the terminal circular orbits are related by $T_2 > T_1$ because r_2 is the larger terminal orbit radius. The difference in mean motion, $n_1 - n_2 > 0$, is the *relative* orbital angular velocity of the target with respect to the vehicle. Because the terminal orbits are circular this value is the correct instantaneous value throughout the motion. The period of this relative motion is called the *synodic period*, S, and is given by

$$S = \frac{2\pi}{n_1 - n_2} = \frac{T_1 T_2}{T_2 - T_1} \qquad (6.26)$$

Data for the orbit periods and synodic periods (with respect to the earth) of several planets is given in Table 6.1. As one would expect from Eq. (6.26), the planets that are nearest the earth have the largest synodic period because the difference in periods is small. By contrast, Pluto moves very little during one year, so its synodic period is only slightly longer than the one year it takes the earth to orbit the sun.

To introduce even more new vocabulary, the synodic period can be interpreted as the time between one *syzygy* and the next. The word *syzygy* means an alignment along a radial line from the sun; it is an impressive sounding word which is worth a lot of points in certain word games.

In the case of the planets, the syzygies relative to the earth are called *conjunctions* and *oppositions*. As shown in Fig. 6.13 the word *opposition* refers to the situation in which the directions from earth to the planet and sun are opposite. This can occur only for *superior* (outer) planets. The word

Impulsive Orbit Transfer

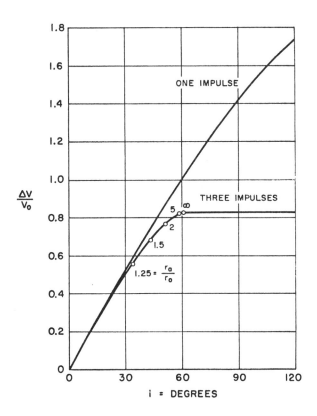

Fig. 6.12 Transfers between Inclined Circular Orbits of Equal Radius (Copyright, 1966, American Institute of Aeronautics and Astronautics; reprinted with permission)

conjunction implies that the directions from earth to the sun and the planet are the same. As shown, the conjunctions for *inferior* (inner) planets are classified as *superior* (farther than the sun) or *inferior* (closer than the sun). The time between two successive oppositions of Mars is simply the synodic period of Mars, and the time between inferior and superior conjunctions of Venus is half the synodic period of Venus.

The condition that assures interception of a target clearly depends on the location of the target relative to the vehicle at the launch time t_1. As shown in Fig. 6.14 the total transfer angle θ must be equal to the sum of the initial phase angle β of the target relative to the vehicle and the angle through

Table 6.1 Synodic Periods

Planet	Orbit Period	Synodic Period (days)
Mercury	88d	116
Venus	224d	584
Mars	687d	778
Jupiter	11.9y	398
Pluto	247y	367

which the target travels in transfer time $t_f = t_2 - t_1$. The intercept condition can then be stated as the required initial phase angle given by:

$$\beta = \theta - n_T\, t_f \qquad (6.27)$$

where n_T is the mean motion of the target. The general condition of Eq. (6.27) can be specialized to the Hohmann transfer by substituting $\theta = \pi$ and t_f equal to the Hohmann transfer time of Eq. (6.5). Two results can then be determined, first for inner-to-outer (r_1 to r_2) orbit interception (for which $n_T = n_2$). In nondimensional form:

$$\beta_{12} = \pi \left[1 - \left[\frac{1+R}{2R}\right]^{3/2}\right] \qquad (6.28)$$

and the second for outer-to-inner (r_2 to r_1) interception (for which $n_T = n_1$):

$$\beta_{21} = \pi \left[1 - \left[\frac{1+R}{2}\right]^{3/2}\right] \qquad (6.29)$$

The reader can verify that, because $R > 1$, the angle β_{12} is positive (i.e., a *lead* angle) and the angle β_{21} is negative, representing a *lag* angle for the target.

Using these relationships one can determine, for example, that for an earth-to-Mars Hohmann transfer, for which $R = 1.524$, the required lead angle of Mars relative to earth is 44°. How often does Mars have that position relative to the earth? The answer is once every synodic period, because that is the period of the relative motion. Referring to Table 6.1, one finds that an earth-to-Mars Hohmann transfer is available every 778 days or 2.13 years.

Impulsive Orbit Transfer 117

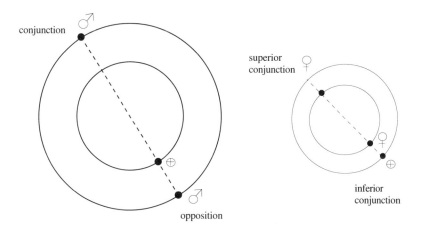

Fig. 6.13 Conjunction and Opposition (with Earth)

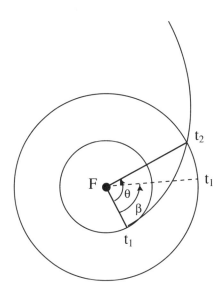

Fig. 6.14 The Intercept Condition

References

6.1 Robbins, H. M., "Analytical Study of the Impulsive Approximation," *AIAA Journal*, **4**, 8, August 1966, pp. 1417–1423.

6.2 Palmore, J., "An Elementary Proof of the Optimality of Hohmann Transfers," *Journal of Guidance, Control and Dynamics*, **7**, 5, September–October 1984, pp. 629–630.

6.3 Lawden, D. F., *Optimal Trajectories for Space Navigation*, Butterworths, London, 1963.

6.4 Edelbaum, T. N., "How Many Impulses?," Paper 66–7, 1966 Aerospace Sciences Meeting, American Institute of Aeronautics and Astronautics (AIAA); also appears in *Astronautics and Aeronautics*, November 1967, pp. 64–69.

6.5 Prussing, J. E., "Simple Proof of the Global Optimality of the Hohmann Transfer," *Journal of Guidance, Control, and Dynamics*, **15**, 4, July-August 1992, pp. 1037-38.

Problems

6.1 a) For a Hohmann transfer between coplanar circular orbits of radii r_1 and r_2, determine expressions for the magnitudes of the velocity changes Δv_1 and Δv_2. Express your answers in dimensionless form $\Delta v_1 / v_{c1}$ and $\Delta v_2 / v_{c1}$ as functions of $R = r_2/r_1$.
b) Obtain a value for $\Delta v_{TOT} / v_{c1}$ in the limit as $R \to \infty$.

6.2 a) For an earth–Mars–Hohmann transfer, calculate Δv_1, Δv_2, and Δv_{TOT} in units of EMOS and km/s.
c) For an exhaust velocity of 0.1 EMOS, determine the ratio of propellant mass to initial mass that is required to supply the necessary Δv_{TOT}.
d) For a total initial mass (prior to the first impulse) of 100 units, using the propellant mass of (c) and a structural coefficient of $\varepsilon = 1/7$, calculate the payload mass m_L (after the second impulse).

6.3 Repeat Prob. 6.2 (a) and (b) for an earth–Jupiter Hohmann transfer with $R = 5.203$.
c) Assuming a specific impulse of $I_{sp} = 450$ s, calculate the required ratio of propellant mass to initial mass.
d) For a total initial mass m_o of 100 units, what is the value of the structural coefficient ε required for a payload mass m_L of 1 unit.

Impulsive Orbit Transfer

6.4 Consider a Hohmann transfer from a 278-km altitude circular orbit about the earth to GEO (geosynchronous earth orbit), which is a circular orbit having a period of one *sidereal day* (relative to the stars) of 23.934 hours, as contrasted with a *solar day* (relative to the direction of the sun) of 24 hours.
a) Determine the required transfer time in hours.
b) Determine the velocity change magnitudes Δv_1 and Δv_2. Include in the second impulse a plane change of 28° by treating $\Delta \mathbf{v}_2$ as a vector which simultaneously circularizes the orbit and makes the plane change.
c) For a value of $I_{sp} = 300$ s, determine the mass ratio Z and the masses m_p, m_s, and m_L for $m_o = 100$ units and $\varepsilon = 1/7$.

6.5 Consider a round-trip earth–Mars–earth mission using Hohmann transfers both ways. Assume coplanar circular terminal orbits of radii 1 au and 1.524 au.
a) Calculate the angles β_{12} and β_{21}.
b) Sketch the planet locations at the launch date at earth and at the arrival date at Mars.
c) How many days after launch from earth does an opposition of Mars occur?
d) Determine the angular separation between two successive oppositions of Mars.
e) Once rendezvous with Mars has occurred, what is the waiting time until a return Hohmann transfer to earth becomes available?

6.6 Repeat Prob. 6.5 for an earth–Jupiter–earth mission with Jupiter at 5.203 au.

6.7* a) Derive Eq. (6.21), which determines the value of the radius ratio R at which the fuel requirement of the bi-parabolic transfer is the same as the Hohmann transfer.
b) Obtain the roots of this cubic equation.

6.8* a) Derive Eq. (6.22), which determines the value of the radius ratio R at which the Hohmann Δv curve achieves its maximum value.
b) Obtain the roots of this cubic equation.

6.9* a) Derive Eq. (6.23), which determines the value of the radius ratio R for which the Hohmann fuel cost is equal to the fuel cost to escape the center of attraction with a single impulse.
b) Obtain the roots of this cubic equation.

7
Interplanetary Mission Analysis

7.1 Introduction

Two aspects of interplanetary mission analysis are treated in this chapter: the *patched conic* method and *planetary flyby* trajectories, also known as *gravity-assist* trajectories.

The patched conic method is a technique for analyzing a complex mission involving a spacecraft and several celestial bodies as a sequence of two-body problems, with one body always being the spacecraft. The basic idea behind this approximate method is that if the spacecraft is sufficiently close to one celestial body, such as the earth, one can, as an approximation, neglect the gravitational forces on the spacecraft due to the sun, moon, and other planets, and analyze the two-body earth-spacecraft problem. The region inside of which this is valid is called the *sphere of influence* of the earth. Each celestial body has such a sphere of influence. If the spacecraft is not inside the sphere of influence of a planet or moon in our solar system, it is considered to be in orbit about the sun.

This concept was briefly mentioned in Chap. 1 as part of the justification for analyzing the two-body problem. By approximating a complex mission as a sequence of two-body problems, one can always use conic orbits to describe the various phases of the mission. As an example, for an earth–Mars mission, the spacecraft begins the mission inside the sphere of influence (SOI) of the earth and exits the sphere of influence using a hyperbolic escape orbit relative to the earth. Once outside the earth's SOI it is in an elliptic orbit about the sun until it becomes close enough to Mars to enter Mars' SOI, after which it is on a hyperbolic approach orbit relative to Mars.

The other topic of this chapter, planetary flyby trajectories, is related to the SOI concept because these encounters occur inside the SOI of the flyby planet. Planetary flyby have been used extensively by interplanetary spacecraft, such as Voyager 1 flying by Jupiter and Saturn, and Voyager 2 flying by Jupiter, Saturn, Uranus, and finally Neptune in 1989. Other examples are ICE (Interplanetary Cometary Explorer), which flew by earth's moon enroute to the comet Giacobinni-Zinner in 1984, and Galileo, which used a VEEGA trajectory, which stands for Venus–Earth–Earth Gravity-Assist, employing a

Interplanetary Mission Analysis

flyby of Venus followed by two flybys of earth prior to reaching Jupiter in 1995.

7.2 Sphere of Influence

To correctly define the SOI, one must be *very* careful, as the discussion that follows will demonstrate. Let us first take the simplistic approach that the spacecraft is within the earth's SOI if the gravitational force on the spacecraft due to the earth is larger than the gravitational force due to the sun. The surface along which the two forces are equal will then be the SOI. To apply this definition one uses Eq. (1.2) and compares

$$\frac{G m_e m_v}{r_{ev}^2} > \frac{G m_s m_v}{r_{sv}^2} \tag{7.1}$$

where the subscript s refers to the sun, e to the earth and v to the vehicle. Equation (7.1) is satisfied if

$$r_{ev} < \left[\frac{m_e}{m_s}\right]^{1/2} r_{sv} \tag{7.2}$$

If we assume the spacecraft is between the earth and the sun ($r_{ev} + r_{sv} = 1$ au $\approx 1.5 \times 10^8$ km), Eq. (7.2) is then satisfied if

$$r_{ev} < \frac{(m_e/m_s)^{1/2}}{1 + (m_e/m_s)^{1/2}} \text{ au} \tag{7.3}$$

which, because $m_s/m_e \approx 3 \times 10^5$, is approximately 2.5×10^5 km. This corresponds to 42 earth radii, or about 70 percent of the distance of 60 earth radii from the earth to the moon.

If one defines this distance to be the boundary of the earth's SOI. the moon would lie outside the SOI and therefore be in orbit about the sun like an asteroid! Since it is an observed fact that the moon is in a nearly circular orbit about the earth, this definition of SOI is obviously incorrect. Actually the earth and moon both orbit their barycenter, which is inside the earth a distance of approximately 0.72 earth radii from the center of the earth.

The correct definition of SOI, due to Laplace in the Eighteenth Century, involves considering the spacecraft to be a satellite of one body, then computing the disturbing acceleration of this motion due to the attraction of the other body. By doing this for each body in turn it is possible to determine, by comparing the ratio of the disturbing acceleration to the central body attraction, which body has the more dominant effect on the spacecraft motion. In order to do this, one must analyze the disturbed relative motion of two bodies.

From Eq. (1.4) one can write an equation similar to Eq. (1.22) but one that includes the effects of all n bodies:

$$\ddot{\mathbf{R}}_1 = \frac{Gm_2}{r_{12}^3} \mathbf{r}_{12} + G \sum_{j=3}^{n} \frac{m_j}{r_{1j}^3} \mathbf{r}_{1j}$$

$$\ddot{\mathbf{R}}_2 = \frac{Gm_1}{r_{12}^3} \mathbf{r}_{21} + G \sum_{j=3}^{n} \frac{m_j}{r_{2j}^3} \mathbf{r}_{2j} \qquad (7.4)$$

Subtracting, and using the fact that $\mathbf{r} = \mathbf{r}_{12} = -\mathbf{r}_{21}$:

$$\ddot{\mathbf{r}} + \frac{\mu}{r^3} \mathbf{r} = -G \sum_{j=3}^{n} m_j \left[\frac{\mathbf{d}_j}{d_j^3} + \frac{\boldsymbol{\rho}_j}{\rho_j^3} \right] \qquad (7.5)$$

where $\mathbf{d}_j$ denotes $-\mathbf{r}_{2j}$ and $\boldsymbol{\rho}_j$ denotes $\mathbf{r}_{1j}$ as depicted in Fig. 7.1.

The term on the right-hand side of Eq. (7.5) is the *disturbing function* which accounts for what are usually called *third body effects*. In the two-body equation of motion Eq. (1.24), this term is equal to 0.

In the application we are dealing with, m_2 is the orbiting spacecraft, the central body m_1 is either a planet or the sun, and the corresponding disturbing third body is either the sun or the planet. In order to emphasize this particular application, let us modify the notation in Eq. (7.5) using subscripts *v* for vehicle, *s* for sun and *p* for planet.

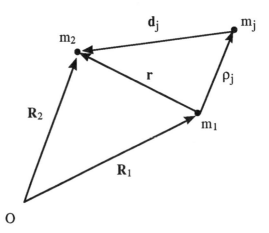

Fig. 7.1 Disturbed Motion

Interplanetary Mission Analysis

For the disturbed motion of the vehicle relative to the planet Eq. (7.5) becomes:

$$\ddot{\mathbf{r}}_{pv} + \frac{G(m_p + m_v)}{r_{pv}^3} \mathbf{r}_{pv} = -Gm_s \left[\frac{\mathbf{r}_{sv}}{r_{sv}^3} - \frac{\mathbf{r}_{sp}}{r_{sp}^3} \right] \tag{7.6}$$

or, simply

$$\ddot{\mathbf{r}}_{pv} - \mathbf{A}_p = \mathbf{P}_s \tag{7.7}$$

where $\mathbf{A}_p$ represents the gravitational acceleration due to the planet and $\mathbf{P}_s$ is the perturbation due to the sun. The position vectors are shown in Fig. 7.2.

Similarly, for the disturbed motion of the vehicle relative to the sun:

$$\ddot{\mathbf{r}}_{sv} + \frac{G(m_s + m_v)}{r_{sv}^3} \mathbf{r}_{sv} = -Gm_p \left[\frac{\mathbf{r}_{pv}}{r_{pv}^3} + \frac{\mathbf{r}_{sp}}{r_{sp}^3} \right] \tag{7.8}$$

or,

$$\ddot{\mathbf{r}}_{sv} - \mathbf{A}_s = \mathbf{P}_p \tag{7.9}$$

where $\mathbf{P}_p$ is the perturbing acceleration due to the planet.

As seen in Eqs. (7.6) and (7.7) when the vehicle is very near the planet, P_s is equal to the difference between two nearly equal vectors and is very small compared to A_p, even though $m_s \gg m_p + m_v$. Also, when the vehicle is far from the planet, P_p is small compared to A_s because m_p is so small in comparison with m_s.

The correct definition of the sphere of influence is that it is the surface along which the magnitudes of the accelerations satisfy

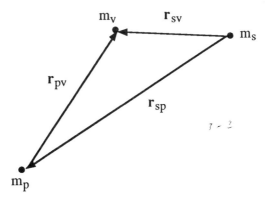

Fig. 7.2 Relative Position Vectors

$$\frac{P_p}{A_s} = \frac{P_s}{A_p} \qquad (7.10)$$

If the left-hand side of Eq. (7.10) is greater than the right-hand side, the spacecraft is inside the SOI of the planet. Note that this contrasts sharply with the previous, incorrect definition, which was simply that the spacecraft was inside the SOI if $A_p/A_s > 1$. The analogous statement using Eq. (7.10) is that $A_p/A_s > P_s/P_p$, where the ratio of the perturbations is a number significantly smaller than unity, resulting in a much larger SOI. For earth the value is approximately 0.15.

Because r_{pv} on the SOI is much smaller than either r_{sv} or r_{sp}, as shown in Battin [7.1] the surface represented by Eq. (7.10) is approximately spherical with center at the planet center and radius given by

$$r_{SOI} \approx \left[\frac{m_p}{m_s}\right]^{2/5} r_{sp} \qquad (7.11)$$

Table 7.1 shows the SOI radius for various planets and for the moon, in which case it is the SOI relative to earth perturbations. Note that the earth SOI is 145 earth radii, as opposed to the 42 earth radii value obtained using the incorrect definition earlier in this chapter. The moon, at 60 earth radii from the earth, is then well within the earth's SOI.

7.3 Patched Conic Method

The *patched conic method* is designed to account for the fact that the planets are not massless, as was the tacit assumption in Chap. 6 when the Hohmann transfer and its extensions were discussed. In that chapter the calculation of the required velocity changes did not consider any masses in the problem other than the spacecraft and the central body.

As determined in the previous section, the earth's SOI radius is approximately 145 earth radii, which is very large compared to the size of the earth. However, on the scale of the solar system this represents a distance of only 6×10^{-3} au, which is only a point on a map of the solar system. This shows the essential idea behind the patched-conic method — namely an interesting application of the *double-think* phenomenon. The SOI is both extremely large and extremely small, depending on the frame of reference in which it is viewed. Relative to the size of the earth it is essentially infinite, while in the solar system it is essentially a point.

Because the SOI is extremely large, essentially infinite relative to the planet size, the velocity relative to the planet exiting the SOI on an escape hyperbola *is considered to be the hyperbolic excess velocity vector*. When this velocity vector is added to the planet's heliocentric velocity, the result is

Table 7.1 Sphere of Influence Radii

Celestial Body	Equatorial Radius (km)	SOI Radius (km)	SOI Radius (body radii)
Mercury	2487	1.13×10^5	45
Venus	6187	6.17×10^5	100
Earth	6378	9.24×10^5	145
Mars	3380	5.74×10^5	170
Jupiter	71370	4.83×10^7	677
Neptune	22320	8.67×10^7	3886
Moon	1738	6.61×10^4	38

the spacecraft's heliocentric velocity on the interplanetary elliptic transfer orbit at the (point-sized) SOI in the solar system. During the elliptic transfer the spacecraft is considered to be under the influence of the sun's gravity only. Upon arrival at the destination planet, the vehicle enters the planet SOI and approaches the planet on an inbound hyperbolic orbit relative to the planet. The same method of patching conics applies here. This demonstrates the basic idea used in applying the patched-conic technique, which will be described more fully shortly. The patched-conic method, as described earlier for interplanetary transfers, does not apply as well for earth–moon transfers because the SOI of the moon lies entirely within the SOI of the earth, so that the earth SOI is never exited during the transfer.

To describe these ideas mathematically, let us employ the notation that $\mathbf{v}_{ba}$ represents the velocity vector of a body a relative to body b. In terms of the three bodies: v = vehicle, p = planet, and s = sun, one can write:

$$\mathbf{v}_{sv} = \mathbf{v}_{sp} + \mathbf{v}_{pv} \tag{7.12}$$

which simply states that by vectorially adding the velocity of the vehicle relative to the planet and the velocity of the planet relative to the sun, one obtains the velocity vector of the vehicle relative to the sun.

To apply Eq. (7.12) at the entrance to or exit from an SOI, one replaces $\mathbf{v}_{pv}$ by the $\mathbf{v}_\infty$ of the planetocentric hyperbolic orbit. Note that this velocity, equal to $\mathbf{v}_{sv} - \mathbf{v}_{sp}$ was interpreted as the required $\Delta \mathbf{v}$ in Chap. 6, where massless planets were assumed. However, when one accounts for the planetary mass, this vector difference in velocities is the hyperbolic excess velocity vector; the actual $\Delta \mathbf{v}$ occurs deep inside the SOI, down near the surface of the planet.

As an application, consider a transfer from earth to another planet. The required $\mathbf{v}_{sv}$ at the earth's SOI is obtained in the general case by first solving Lambert's problem as discussed in Chap. 4. Then, by subtracting the known value of $\mathbf{v}_{sp}$, one obtains the required $\mathbf{v}_{\infty e}$, which represents the hyperbolic excess speed on the earth escape hyperbola. From this the actual required $\Delta \mathbf{v}$ can be determined.

As an illustrative example let us consider the earth–Venus Hohmann transfer treated in Kaplan [7.2]. The mission objective is to take a spacecraft from a 200 km altitude circular parking orbit about the earth to a point 500 km from the surface of Venus. What happens after that does not concern us here and depends on whether the spacecraft will fly by Venus, use a retro thrust to become a Venus orbiter, or plunge into the atmosphere. Our example ends at the 500 km altitude periapse on the hyperbolic approach orbit to Venus. All the conic orbits in this patched-conic analysis are assumed to lie in the same plane, for simplicity, but a three-dimensional analysis could be made using the same technique.

The required excess velocity for the earth escape hyperbola is, for reasons discussed earlier, precisely the same as the Δv for a massless planet, which for a Hohmann transfer is the subject of Prob. 6.1. For $R = r_2/r_1 = 1$ au/0.723 au = 1.383, the value of $\Delta v_2/v_{c1}$ is found to be 0.0703. Note that Δv_2 is used because the earth is the *larger* planetary orbit; recall that by definition $r_2 > r_1$. Thus v_{c1} is Venus' orbital speed relative to the sun of 1.18 EMOS, resulting in a Δv_2 value of 0.083 EMOS or 2.48 km/s. Figure 7.3 shows the direction of this $\mathbf{v}_{\infty e}$ for the earth escape hyperbola. The vehicle exits the earth SOI via the "back door" (i.e. via the trailing edge of the SOI in relation to its motion around the sun). This is, of course, due to the fact that the required aphelion speed on the Hohmann transfer ellipse is less than earth's orbital speed. The vehicle must be traveling slower than the earth in order to fall closer to the sun, where Venus lives.

At Venus arrival the magnitude of the $\mathbf{v}_{\infty v}$ at Venus SOI is determined by calculating Δv_1 for a massless planet to be 0.0910 EMOS or 2.71 km/s,

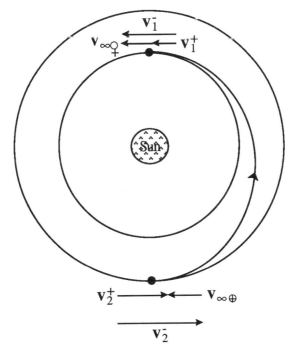

Fig. 7.3 Earth–Venus Hohmann Transfer

with a direction shown in Fig. 7.3. The vehicle enters the back door of Venus' SOI because the vehicle heliocentric arrival velocity is greater than the planet's heliocentric orbital velocity.

Meanwhile, back at the earth SOI, the $v_{\infty e}$ magnitude of 2.48 km/s along with the perigee altitude of 200 km determines the earth escape hyperbola. The eccentricity and turn angle are determined using the formulas of Prob. 1.14 to be $e = 1.10$ and $\delta/2 = 65.2°$. The perigee speed on the earth escape hyperbola, v_{pe}, is determined by

$$v_{pe}^2 = \frac{2\mu_e}{r_{pe}} + v_{\infty e}^2 \qquad (7.13)$$

where r_{pe} is the perigee radius of the earth escape hyperbola, equal to 6378 + 200 = 6578 km. The result is $v_{pe} = 11.28$ km/s. The *actual* Δv required, then, is the difference between this speed and the circular parking orbit speed at 200 km altitude, which is 7.78 km/s. The result of the calculation is that the actual $\Delta v_e = 3.5$ km/s, which is significantly different from the massless planet value of 2.48 km/s. In this instance the actual velocity change is larger than the velocity change for a massless planet, but there are also situations where the actual velocity change is smaller than the massless planet value.

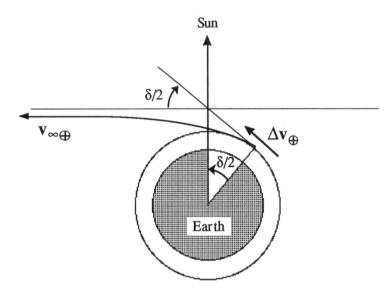

Fig. 7.4 Earth Escape Δv Geometry

Figure 7.4 shows the geometry of the earth escape hyperbola relative to the earth–sun line in Fig. 7.3. The point in the circular parking orbit at which the actual Δv is tangentially applied must be δ/2 = 65.2° prior to crossing the earth–sun line in order that the $\mathbf{v}_{\infty e}$ have the correct direction as shown.

At the Venus end of the mission the periapse altitude of 500 km, when added to the planet radius of 6187 km yields $e = 1.15$ and δ/2 = 60.3°, along with an aiming radius Δ = 25,270 km or about 4.1 Venus radii. Note that, as shown in Fig. 7.5, the spacecraft could approach Venus on either the sunlit side or the dark side with the same $\mathbf{v}_\infty$ and Δ by a very slight crosstrack velocity adjustment very far from the planet. The choice would depend on the details of the mission after arrival, which we are not considering in this illustrative example.

7.4 Velocity Change from Circular to Hyperbolic Orbit

In the previous section the tangential velocity change from a circular parking orbit to a hyperbolic orbit having a given v_∞ was calculated. It is of interest to perform a general analysis of the coplanar impulsive circle to (or from) hyperbola maneuver. Because the velocity change is tangential, the periapse of the hyperbolic orbit is equal to the circular orbit radius. On the hyperbola

$$v_p^2 = 2v_c^2 + v_\infty^2 \qquad (7.14)$$

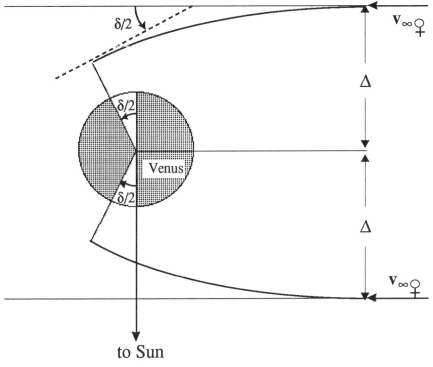

Fig. 7.5 Venus Arrival Geometry

and $\Delta v = v_p - v_c$, resulting in

$$\frac{\Delta v}{v_c} = \left[2 + \left(\frac{v_\infty}{v_c}\right)^2\right]^{1/2} - 1 \qquad (7.15)$$

Figure 7.6 shows the plot of the velocity change in Eq. (7.15). As seen in the figure the comparison between the actual Δv and the velocity change for a massless planet (equal to v_∞) is very simple. For $0 < v_\infty/v_c < \frac{1}{2}$ the actual $\Delta v > v_\infty$, whereas for $v_\infty/v_c > \frac{1}{2}$, $\Delta v < v_\infty$. In the earth–Venus Hohmann transfer in Sect. 7.3, at earth $v_\infty = 2.48$ km/s, $v_c = 7.78$ km/s, resulting in $v_\infty/v_c = 0.32 < 0.5$, and the $\Delta v = 3.50$ km/s $> v_\infty$.

7.5 Planetary Flyby (Gravity-Assist) Trajectories

A planetary *flyby trajectory*, also called a *gravity-assist maneuver* and a *swingby trajectory*, is useful in interplanetary missions to obtain a velocity change without expending propellant. This "free" velocity change is provided by the gravitational field of the flyby planet. It can be used to lower the propellant Δv cost of a mission and thereby increase the the payload mass. It can also be used during a return transfer to earth to decrease the

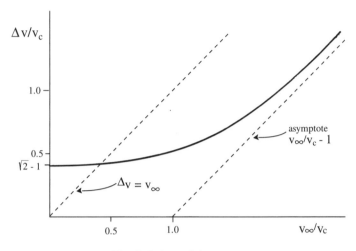

Fig. 7.6 Actual Δv vs. v_∞

reentry speed upon reaching the earth's atmosphere. In a more passive application it can be used to simply obtain a close-up image of a planet or moon.

Figure 7.7 shows the hyperbolic flyby trajectory *relative to the planet*. The vectors $\mathbf{v}_{\infty I}$ and $\mathbf{v}_{\infty O}$ are the respective inbound and outbound hyperbolic excess vectors entering and exiting the flyby planet SOI. The direction of the planet's heliocentric velocity $\mathbf{v}_{sp}$ is also shown.

Figure 7.8 demonstrates the basic velocity vector geometry for a planetary flyby. As in Eq. (7.12) the heliocentric velocity of the vehicle $\mathbf{v}_{sv}$ immediately before and after the flyby (i.e., entering and exiting the SOI) is obtained by vectorially adding the corresponding $\mathbf{v}_\infty$, which is relative to the planet, to the planet's heliocentric velocity $\mathbf{v}_{sp}$. The two values of the heliocentric velocity vector shown are $\mathbf{v}_{svI}$ and $\mathbf{v}_{svO}$, denoting inbound (before) and outbound (after) values. The vector difference between them is the velocity change provided by the flyby planet:

$$\Delta \mathbf{v}_{FB} = \mathbf{v}_{svO} - \mathbf{v}_{svI} = \mathbf{v}_{\infty O} - \mathbf{v}_{\infty I} \tag{7.16}$$

Note that the assumption has been made that the values of $\mathbf{v}_{sp}$ and r_{sp} have not changed during the flyby. This is similar to the impulsive approximation of Chap. 6 and is based on the fact that the duration of the flyby maneuver is very short compared with the period of the flyby planet.

The flyby trajectory depicted in Figs. 7.7 and 7.8 results in an increase in the heliocentric speed of the vehicle: $v_{svO} > v_{svI}$, due to the fact that the

Interplanetary Mission Analysis 131

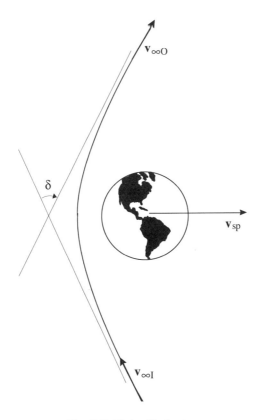

Fig. 7.7 Flyby Trajectory

vehicle *passed behind the flyby planet* in its motion about the sun. This brings up an interesting point: Because the the heliocentric kinetic energy of the vehicle has increased and the potential energy has remained unchanged (because r_{sp} is essentially constant during the maneuver) the total heliocentric energy of the vehicle has increased. The reader may wish to contemplate the question: Where does this energy come from?

In the general case the periapse of the flyby hyperbola will not always lie behind the planet as shown in Fig. 7.7, but may well lie in front of the planet or at other locations. By examining Fig. 7.8, it is easy to see that, in particular, a flyby in front of the planet would cause a decrease in the vehicle's heliocentric speed.

By combining Eq. (7.16) for the Δv_{FB} with the patched- conic technique to determine $v_{\infty I}$, one can determine the value of v_{svO}, which along

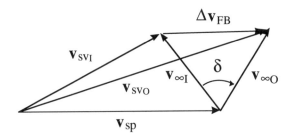

Fig. 7.8 Flyby Velocity Geometry

with r_{sv} (which is same as r_{sp} because the SOI is a point in the solar system), completely determined the heliocentric orbital elements after the flyby, as discussed in Sec. 3.3. Because the Δv_{FB} magnitude can be appreciable for a large planet such as Jupiter, the heliocentric orbit can experience a significant discontinuity in the velocity, resulting in a "corner" on the trajectory as shown in Fig. 7.9, just as if a thrust impulse were applied. In the case of Voyagers I and II, the increase in heliocentric speed due to the Jupiter flyby was enough to place these spacecraft on *hyperbolic trajectories relative to the sun*. Both spacecraft are now on escape orbits from our solar system, venturing out into other parts of our galaxy.

A formula can be developed for the magnitude of the flyby velocity change, Δv_{FB}, in terms of the properties of the flyby hyperbola, such as periapse radius and excess speed. To do this note that the magnitudes $v_{\infty I}$ and $v_{\infty O}$ are equal even though their directions are not the same. This is simply because energy relative to the planet is conserved on a flyby hyperbola if no forces other than planetary gravity act. If a thrust maneuver near periapse is made using a rocket engine, this would not be the case, but let us consider a totally passive, freefall trajectory. The effect of thrust is considered in Prob. 7.6. Referring to Fig. 7.8 one can write

$$\Delta v_{FB} = 2 v_\infty \sin \frac{\delta}{2} \qquad (7.17)$$

where v_∞ is the common magnitude of both the inbound and outbound excess velocity vectors. Now we have found another application of Prob. 1.14 and we can write Eq. (7.17) in nondimensional form as

$$\frac{\Delta v_{FB}}{v_s} = \frac{2 v_\infty / v_s}{1 + (\frac{v_\infty}{v_s})^2 (\frac{r_p}{r_s})} \qquad (7.18)$$

Interplanetary Mission Analysis 133

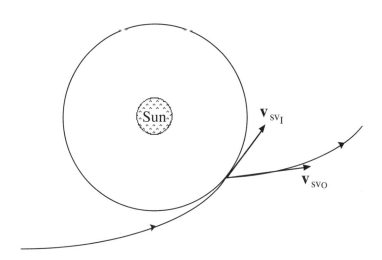

Fig. 7.9 Discontinuity in Velocity due to Flyby

where, as before, r_s is the planet radius and v_s is the circular orbit speed at the surface of the planet. Recall that $r_s v_s^2$ is simply equal to the gravitational constant μ of the planet (Prob. 1.14).

To apply Eq. (7.18) the value of v_∞ is determined by the heliocentric arrival velocity of the vehicle at the flyby planet, v_{svI}, which is determined by solving Lambert's problem. The specific value of v_∞ comes from the patched-conic analysis. The other variable in Eq. (7.18) is the periapse radius r_p of the flyby hyperbola. This is not determined, but free to be chosen by the selection of an aiming radius Δ. This can be varied over a wide range by a very small crosstrack velocity change out near the SOI when the vehicle is inbound to the planet. The value of r_p and Δ are related as in Prob. 1.14.

An obvious constraint to be satisfied in a successful flyby is $r_p \geq r_s$, otherwise the flyby would abruptly terminate. *It is important to emphasize*, as illustrated in Fig. 7.8, that $v_{svO} \neq v_{svI} + \Delta v_{FB}$ (i.e., one cannot simply add the velocity magnitudes). A vectorial addition must be made to determine the outbound heliocentric speed and the law of cosines or law of sines is required to solve for v_{svO}.

Table 7.2 Circular Orbit Speed at Surface of Planet

Planet	v_s (km/s)	v_s (EMOS)
Mercury	2.996	0.101
Venus	7.256	0.244
earth	7.905	0.265
Mars	3.569	0.120
Jupiter	42.15	1.415
Saturn	25.07	0.841
Uranus	15.73	0.528
Neptune	17.58	0.590
Pluto	0.49	0.017
Sun	436.7	14.66
Moon	1.680	0.056

Table 7.2 lists the values of v_s for the various planets along with the values for the sun and earth's moon for use in Eq. (7.18).

7.6 Flyby Following a Hohmann Transfer

The special case of a flyby that follows a heliocentric Hohmann transfer is of interest because some special characteristics can be discovered. Figure 7.10 shows the flyby vector velocities after a Hohmann transfer from earth to an inner planet in the solar system. For the case shown the periapse of the flyby hyperbola is on the dark side of the planet, resulting in a negative heliocentric flight path angle after the flyby. A periapse on the sunlit side would have resulted in a positive flight path angle. As shown, the heliocentric velocity of the vehicle is *always decreased* by the flyby maneuver, due to the fact that $\mathbf{v}_{svl}$ is parallel to $\mathbf{v}_{sp}$ and larger in magnitude. As a result, the vector $\mathbf{v}_{\infty l}$ is in the same direction as $\mathbf{v}_{sp}$, indicating that the vehicle enters the back door of the SOI.

Figure 7.11 shows the analogous situation following a Hohmann transfer to an outer planet. In this case $\mathbf{v}_{svl}$ is again parallel to $\mathbf{v}_{sp}$, but smaller in magnitude. As a result, the vector $\mathbf{v}_{\infty l}$ is directed opposite to $\mathbf{v}_{sp}$, indicating that the vehicle enters the front door of the SOI. The planet

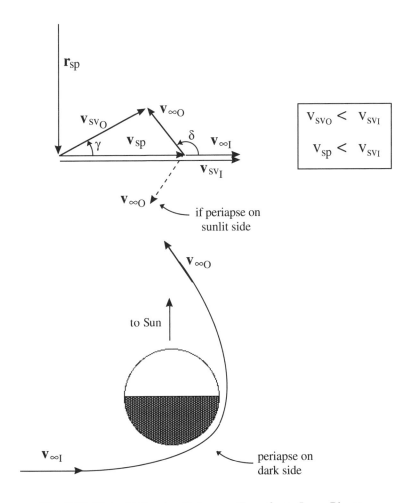

Fig. 7.10 Flyby Following Hohmann Transfer to Inner Planet

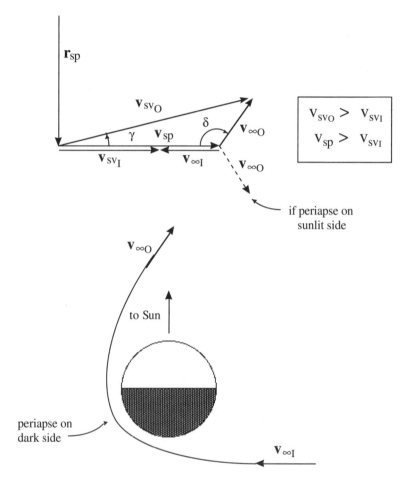

Fig. 7.11 Flyby Following Hohmann Transfer to Outer Planet

overtakes the vehicle in its motion around the sun. Relative to an observer on the planet, the vehicle approaches the planet from a point ahead of the planet.

The flyby maneuver in this case *always increases* the heliocentric speed of the vehicle, as shown in Fig. 7.11. For a periapse on the dark side, the heliocentric flight path angle will be negative after the flyby and for a sunlit side periapse it will be positive.

All of the flyby maneuvers considered assume a free-fall trajectory by the planet. Another possibility is to apply a thrust impulse during the flyby. Prob. 7.6 considers the use of a thrust impulse at the periapse of the flyby hyperbola.

References

7.1 Battin, R. H., *An Introduction to the Mathematics and Methods of Astrodynamics.* AIAA Education Series, 1987.

7.2 Kaplan, M. H., *Modern Spacecraft Dynamics and Control.* John Wiley and Sons, 1976.

Problems

7.1 Determine the erroneous value of r_{ev} in Eq. (7.2) if the earth is directly between the spacecraft and the sun.

7.2 Consider an earth–Mars Hohmann transfer using the patched-conic technique.
a) Calculate the actual Δv magnitudes (in EMOS) to tangentially depart a posigrade circular parking orbit about the earth of radius 1.1 earth radii and terminate in a posigrade circular orbit about Mars of radius 1.3 Mars radii. Assume all orbits are coplanar.
b) Determine the required angle between the departure point in the earth parking orbit and the sun–earth line. Sketch the geometry.
c) Determine the required aiming radius at Mars in units of Mars radii.

7.3* Using the patched-conic technique, compare the Δv cost of a one-impulse (intercept) Hohmann transfer to Jupiter with the Δv cost to escape the solar system with an impulsive thrust at the earth. Specifically:
a) Determine the Δv from a 1.1 earth radii circular parking orbit about the earth that will place the spacecraft on the Hohmann ellipse to Jupiter.
b) Determine the Δv from a 1.1 earth radii parking orbit about the

earth that will place the spacecraft on a parabolic heliocentric orbit at the earth's SOI.

c) Compare the two Δv costs.

d) Compare the final masses for $I_{sp} = 400$ s.

7.4 For a hyperbolic flyby of a planet,

a) Determine the values of periapse radius r_p and hyperbolic excess speed v_∞ that yield the *maximum possible* magnitude of Δv_{FB}. Express your answer for r_p in terms of the planet radius r_s and include the constraint $r_p \geq r_s$.

b) Determine this maximum value in terms of v_s and determine numerical values for the corresponding turn angle and eccentricity.

7.5 Using the patched-conic technique, consider an approximate *Voyager* trajectory, namely a Hohmann transfer to Jupiter, followed by a hyperbolic flyby of Jupiter. Assume coplanar, circular orbits for earth and Jupiter of radii 1 au and 5.2 au and choose the periapse radius of the flyby hyperbola to maximize Δv_{FB}.

a) Calculate the speed of the spacecraft relative to the sun after the flyby and compare to escape speed at 5.2 au.

b) Calculate the value of true anomaly on the heliocentric orbit immediately after the flyby, assuming the periapse of the flyby hyperbola is on the sunlit side of Jupiter.

c) Calculate the transfer angle on the heliocentric orbit after the flyby between Jupiter and the crossing of Saturn's orbit.

7.6 Consider a first-order variation δv_p in the periapse speed on a flyby hyperbola due to an impulsive thrust at periapse. Assume the velocity change is parallel to the periapse velocity. Derive an expression for the relative or "marginal" change in hyperbolic excess speed $\delta v_\infty / v_\infty$ in terms of the relative change in periapse speed $\delta v_p / v_p$. Remember that the periapse radius is unchanged by the thrust impulse at periapse.

8
Linear Orbit Theory

8.1 Introduction

For the two-body problem in a nonlinear, inverse-square gravitational force field we have seen in previous chapters that much of the analysis can be performed in closed form in terms of simple conic orbits. However, some calculations, such as the iterative solution of Kepler's equation and Lambert's problem, and the determination of impulsive velocity changes, are still somewhat complicated. For this reason a simpler, approximate mathematical model is useful. In this chapter we will derive, analyze, and apply linearized equations that approximately describe the motion of a vehicle relative to a neighboring reference orbit. There are many applications of linear orbit theory, including midcourse guidance, close-range interception and rendezvous, and satellite evasive maneuvers. The great advantage in dealing with linear equations of motion is that superposition applies. The total motion due to thrust and gravity can be obtained by superposition of the separate effects.

8.2 Linearization of the Equations of Motion

Let us first approach the problem in terms of a general gravitational field, for which the equation of motion of our vehicle is

$$\ddot{\mathbf{r}} = \mathbf{g}(\mathbf{r}) + \mathbf{\Gamma} \tag{8.1}$$

where $\mathbf{\Gamma}$ is the thrust acceleration vector, equal to $-b\,\mathbf{c}/m$ from Chap. 5. Figure 8.1 depicts the situation where we are considering: $\mathbf{r}(t)$ represents the orbit of the vehicle, and $\mathbf{r}^*(t)$ represents the orbit of a reference body (for example, a target such as a space station) in a *known* reference orbit. This means that $\mathbf{r}^*(t)$ is known either analytically or as a result of a numerical integration.

The vector difference between the vehicle position and the target position is shown as the *relative position vector* $\delta\mathbf{r}(t)$:

$$\delta\mathbf{r}(t) = \mathbf{r}(t) - \mathbf{r}^*(t) \tag{8.2}$$

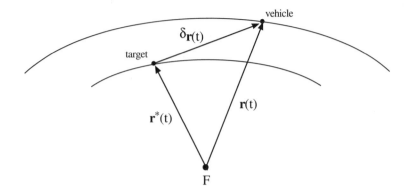

Fig. 8.1 The Relative Position Vector

This difference, or variation, in the radius vector is sometimes referred to as a *contemporaneous* variation, because the variation is defined at a common value of time ($\delta t \equiv 0$).

Our goal is then to determine a simple, approximate equation describing $\delta \mathbf{r}(t)$. The solution to this simple equation can then be added to the known $\mathbf{r}^*(t)$ to determine an approximate value for $\mathbf{r}(t)$ [Eq. (8.2)]. In the case of interception or rendezvous with the target, it is only the *relative* motion described by $\delta \mathbf{r}(t)$ that we are concerned with; an interception occurs when $\delta \mathbf{r} = 0$.

Substituting $\mathbf{r} = \mathbf{r}^* + \delta \mathbf{r}$ into Eq. (8.1) yields

$$\ddot{\mathbf{r}} = \ddot{\mathbf{r}}^* + \delta \ddot{\mathbf{r}} = \mathbf{g}(\mathbf{r}^* + \delta \mathbf{r}) + \boldsymbol{\Gamma} \tag{8.3}$$

We now expand the vector $\mathbf{g}$ in a Taylor series about the reference orbit $\mathbf{r}^*(t)$:

$$\mathbf{g}(\mathbf{r}^* + \delta \mathbf{r}) = \mathbf{g}(\mathbf{r}^*) + \frac{\partial \mathbf{g}(\mathbf{r}^*)}{\partial \mathbf{r}^*} \delta \mathbf{r} + \mathbf{b} \tag{8.4}$$

Linear Orbit Theory

where the vector **b** represents second- and higher-order terms. Note that the partial derivative of the gravitational acceleration vector with respect to the position vector is evaluated along the reference orbit $\mathbf{r}^*(t)$. This partial derivative is a 3×3 matrix $G(\mathbf{r})$, which is called the *gravity gradient matrix*, because it represents the gradient of the gravitational acceleration **g**. Each element of the matrix G is given by

$$G_{ij} = \frac{\partial g_i}{\partial r_j} \tag{8.5}$$

Since the gravitational acceleration **g** is derivable as the negative of the gradient of a scalar potential energy function V (i.e., a conservative force field), the matrix G is *symmetric* ($G_{ji} = G_{ij}$) because

$$G_{ij} = \frac{\partial g_i}{\partial r_j} = -\frac{\partial^2 V}{\partial r_i \partial r_j} \tag{8.6}$$

The second- and higher-order terms denoted by **b** in the expansion of Eq. (8.4) have the form

$$b_i = \frac{1}{2!} \frac{\partial^2 g_i}{\partial r_j \partial r_k} \delta r_j \delta r_k + \frac{1}{3!} \frac{\partial^3 g_i}{\partial r_j \partial r_k \partial r_m} \delta r_j \delta r_k \delta r_m + \cdots \tag{8.7}$$

where the partial derivatives are evaluated along $\mathbf{r}^*$ and the summation convention for repeated indices is implied. If the magnitude $|\delta\mathbf{r}|$ is small compared to the reference value $|\mathbf{r}^*|$, then $|\delta r_i/r_i^*| \ll 1$ and the second- and higher-order terms comprising **b** can be neglected. As an example, consider LEO, for which r^* is approximately 6500 km. A position variation δr of 65 km yields $\delta r/r^*$ of 0.01. For GEO, $r^* = 42{,}164$ km, so a position variation of 422 km represents 1 percent of the reference value.

If one substitutes Eq. (8.4) into (8.3) and utilizes the fact that the reference orbit itself $\mathbf{r}^*(t)$ also satisfies the equation of motion (8.1), then $\ddot{\mathbf{r}}^* = \mathbf{g}(\mathbf{r}^*)$ and Eq. (8.3) reduces to

$$\delta\ddot{\mathbf{r}} = \frac{\partial \mathbf{g}(\mathbf{r}^*)}{\partial \mathbf{r}^*} \delta\mathbf{r} + \Gamma = G(\mathbf{r}^*)\delta\mathbf{r} + \Gamma \tag{8.8}$$

which is a *linear* differential equation for $\delta\mathbf{r}$.

As a specific application, let us consider the inverse-square gravitational force. In this case the gravity gradient matrix of Eq. (8.5) is found to be [Prob. 8.6]:

$$G(\mathbf{r}) = \frac{\mu}{r^5} (3\mathbf{r}\mathbf{r}^T - r^2 I_3) \tag{8.9}$$

where I_3 is the 3×3 identity matrix.

To explicitly evaluate the elements of the gravity gradient matrix for the inverse-square field, it is convenient to express the vector components in a coordinate frame that rotates with the reference radius vector $\mathbf{r}^*$. In such a frame $\mathbf{r}^*$ has the simple form:

$$\mathbf{r}^* = \begin{bmatrix} r^* \\ 0 \\ 0 \end{bmatrix} \tag{8.10}$$

and $G(\mathbf{r}^*)$ from Eq. (8.9) is the diagonal matrix:

$$G(\mathbf{r}^*) = \frac{\mu}{r^{*3}} \begin{bmatrix} 2 & 0 & 0 \\ 0 & -1 & 0 \\ 0 & 0 & -1 \end{bmatrix} \tag{8.11}$$

The case we will investigate in detail is for a *circular* reference orbit, for which $G(\mathbf{r}^*)$ in Eq. (8.11) is a *constant* matrix, resulting in a *linear, homogeneous* (for $\Gamma = 0$) *constant-coefficient* differential equation for $\delta \mathbf{r}(t)$, which can be solved in closed form.

8.3 The Hill–Clohessy–Wiltshire (CW) Equations

As mentioned previously, a coordinate frame that rotates with the reference radius vector $\mathbf{r}^*$ is a convenient one in which to express vector components. A specific frame of this type that is commonly used is the *local vertical* coordinate frame shown in Fig. 8.2, sometimes referred to as the CW frame after Clohessy–Wiltshire [8.1], although the use of this coordinate frame was first introduced by Hill [8.2]. This frame is also discussed in [8.3]. The origin of this frame is at the reference point (target) in the reference orbit, with the x-

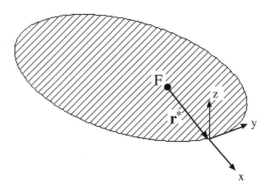

Fig. 8.2 The CW Frame

Linear Orbit Theory

axis directed radially outward along the local vertical, the y-axis along the direction of motion, and the z-axis normal to the reference orbit plane.

The one remaining step in obtaining the final linearized equations of relative motion for a circular reference orbit is to transform the acceleration $\ddot{\delta r}$ of Eq.(8.8), which is relative to an observer in an inertial frame, to the acceleration $(\ddot{\delta r})_R$, which is relative to a rotating observer fixed in the CW frame. This is a natural frame of reference to use because the "down" direction to the center of the planet is along the negative x-axis, which is fixed in the CW frame. The well-known transformation to obtain the acceleration relative to the rotating CW frame is

$$(\ddot{\delta r})_R = \ddot{\delta r} - 2\omega \times (\dot{\delta r})_R - \dot{\omega} \times \delta r - \omega \times (\omega \times \delta r) \tag{8.12}$$

where ω represents the angular velocity of the CW frame with respect to an inertial frame, which is simply the orbital angular velocity in the reference orbit. In Eq. (8.12) the second term on the right-hand side is the Coriolis acceleration and the last term is the centrifugal acceleration. In xyz components:

$$\delta r = \begin{bmatrix} x \\ y \\ z \end{bmatrix} \quad \text{and} \quad \omega = \begin{bmatrix} 0 \\ 0 \\ n \end{bmatrix} \tag{8.13}$$

where n is the mean motion of the circular reference orbit.

The components of the vector derivatives $(\dot{\delta r})_R$ and $(\ddot{\delta r})_R$ in the CW frame are easily obtained because the frame in which the vector components are expressed and the frame with respect to which the derivative is being taken are the same. In this case the components of the vector derivative are simply the derivatives of the individual vector components:

$$(\dot{\delta r})_R = \begin{bmatrix} \dot{x} \\ \dot{y} \\ \dot{z} \end{bmatrix} \quad \text{and} \quad (\ddot{\delta r})_R = \begin{bmatrix} \ddot{x} \\ \ddot{y} \\ \ddot{z} \end{bmatrix} \tag{8.14}$$

Substituting Eqs. (8.11), (8.13) and (8.14) into Eq. (8.12) and then into (8.8) yields

$$\ddot{x} = 3n^2 x + 2n\dot{y} + \Gamma_x \tag{8.15a}$$

$$\ddot{y} = -2n\dot{x} + \Gamma_y \tag{8.15b}$$

$$\ddot{z} = -n^2 z + \Gamma_z \tag{8.15c}$$

where the fact that $\mu/r^{*3} = n^2$ has been utilized.

Equations (8.15) are the Hill–Clohessy–Wiltshire equations in component form, commonly referred to as the CW equations. Clohessy and Wiltshire rediscovered these equations in [8.1] in a study of the motion of a vehicle relative to a satellite in earth orbit. Hill's original study [8.2] in the nineteenth century described the motion of the moon relative to the earth. These two applications are quite different, but they both consider small displacements relative to a known reference motion. In Hill's application the distance of the moon from the earth is small compared to the distance of the earth from the sun. In the Clohessy–Wiltshire application, the distance of the vehicle from an earth-orbiting satellite is small compared to the distance from the satellite to the center of the earth.

These equations can also be expressed in matrix form as:

$$\begin{bmatrix} \ddot{x} \\ \ddot{y} \\ \ddot{z} \end{bmatrix} = A \begin{bmatrix} x \\ y \\ z \end{bmatrix} + B \begin{bmatrix} \dot{x} \\ \dot{y} \\ \dot{z} \end{bmatrix} + \begin{bmatrix} \Gamma_x \\ \Gamma_y \\ \Gamma_z \end{bmatrix} \quad (8.16a)$$

where

$$A = \begin{bmatrix} 3n^2 & 0 & 0 \\ 0 & 0 & 0 \\ 0 & 0 & -n^2 \end{bmatrix} \quad \text{and} \quad B = \begin{bmatrix} 0 & 2n & 0 \\ -2n & 0 & 0 \\ 0 & 0 & 0 \end{bmatrix} \quad (8.16b)$$

8.4 The Solution of the CW Equations

The linear, constant-coefficient differential equations given by Eq. (8.15) are solvable, but it is instructive to first analyze them. One interesting property is that, although the equations describing the in-plane (x-y) motion are coupled, the out-of-plane (z) motion is uncoupled from the in-plane motion.

Another property is that there are velocity-dependent terms $2n\dot{y}$ and $2n\dot{x}$ in these equations, whose coefficients are elements of the B matrix in Eq. (8.16b) and represent damping in the system. However, the damping is of a special type known as *gyroscopic damping*, for which $\ddot{x}$ depends on $\dot{y}$, but not $\dot{x}$, and $\ddot{y}$ depends on $\dot{x}$ but not $\dot{y}$. Damping of this type is nondissipative (total energy is conserved) and is present only because we are describing the motion in a rotating coordinate frame.

Before actually solving the CW Eqs. (8.15) for the unthrusted motion of a vehicle in the neighborhood of a circular orbit, it is instructive to analyze two special cases for the in-plane (xy) motion. Understanding these special cases will aid in understanding what the general solutions for $x(t)$ and $y(t)$ are describing.

First, let us investigate whether *constant* values of x and y will satisfy the CW equations. In other words, is there a solution that is stationary when viewed in the rotating frame. Substituting $x(t) = c_1$ and $y(t) = c_2$, for which $\dot{x} = \ddot{x} = \dot{y} = \ddot{y} = 0$, into the x and y component CW equations for unthrusted motion ($\Gamma_x = \Gamma_y = 0$) yields the solution:

$$3n^2 c_1 = 0$$

which tells us that the constant c_1 must be 0, but the value of c_2 is *arbitrary*. This stationary solution in the CW frame then has the form shown in Fig. 8.3.

If one contemplates how this motion appears in an inertial frame, it is evident that the vehicle in this special case is orbiting in the same circular orbit as the target at the origin, but at a (small) fixed distance c_2 ahead of (for $c_2 > 0$) or behind (for $c_2 < 0$) the target. To an observer fixed in the space station (target) and rotating with its radius vector, the vehicle appears to be at rest.

This is a good spot to discuss the interpretation of the position components along our axis directions. The reader has most likely assumed that the CW frame is a cartesian frame, in which case the constant c_2 described earlier must be quite small in magnitude compared to r^* because the (straight) axis direction approximates the (curved) circular orbit of the target only for small deviations away from the origin. Because this is true for variations out of the target orbit plane as well as for in-plane variations, the linear

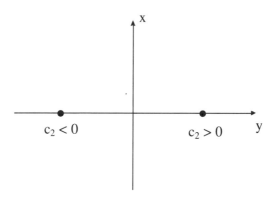

Fig. 8.3 Stationary Solution in the CW Frame

range for the cartesian frame is a sphere of radius much less than r^*, centered at the origin of the CW frame.

Another alternative is to utilize *cylindrical* coordinates, for which the in-plane coordinates are the polar coordinates $x = \delta r$ and $y = r^* \delta\theta$, and the out-of-plane coordinate z is identical to the cartesian coordinate. The advantage of cylindrical coordinates is that the gravitational acceleration is *independent of the longitude* θ, which varies along the circular target orbit. This is true for any central force field.

The CW equations have *exactly the same form* using polar coordinates for x and y and the linear range now becomes a *torus about the circular target orbit*, because only x and z need to be small compared to r^* [8.4]. As an example, the constant c_2 in our special case discussed earlier is now the difference in longitude between the vehicle and the target. The CW equations provide the correct solution even for large values of c_2 because the y-axis now lies *along* the circular target orbit.

Another special case of interest is $x(t) = c_1$ and $\dot{y}(t) = c_3$ (for which $\dot{x} = \ddot{x} = \ddot{y} = 0$). The CW equations for this special case are satisfied if

$$3n^2 c_1 + 2nc_3 = 0 \tag{8.17}$$

which implies that $c_3 = -3nc_1/2$. Examples of this solution in the CW frame are shown in Fig. 8.4. The constant value (c_3) of $\dot{y}$ is negative for positive c_1 and vice versa. In cylindrical coordinates a constant value of c_1 represents a constant difference in the radius ($\delta r = c_1$). This solution then represents motion in a coplanar circular orbit of radius slightly larger ($c_1 > 0$) or smaller ($c_1 < 0$) than the target orbit. The constant $\dot{y} = c_3$ of opposite sign to c_1 merely represents the fact that a larger radius circular orbit will have a larger period and thus a smaller mean motion than the target orbit. A vehicle in this larger orbit will fall behind the target at a constant rate. Likewise, a vehicle in a smaller circular orbit will gain in longitude relative to the target at a constant rate. An observer in the target would see the relative motion depicted in Fig. 8.4 for these special cases.

Returning to the consideration of the general solution of the CW Eqs. (8.15), let us first turn our attention to the out-of-plane motion described by $\ddot{z} + n^2 z = 0$. This undamped linear oscillator has the well-known solution

$$z(t) = z_o \cos nt + \frac{\dot{z}_o}{n} \sin nt \tag{8.18a}$$

$$\dot{z}(t) = -z_o n \sin nt + \dot{z}_o \cos nt \tag{8.18b}$$

where z_o and $\dot{z}_o$ are the respective arbitrary initial conditions on out-of-plane displacement and velocity relative to the target.

Linear Orbit Theory

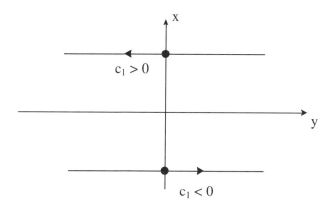

Fig. 8.4 Neighboring Circular Orbits in the CW Frame

Let us consider the particular case $z_o = c_4$ and $\dot{z}_o = 0$ in order to help understand in what way this harmonic motion of angular frequency n (Period $T = \frac{2\pi}{n}$) describes a neighboring orbit. At the initial time $t = 0$ the vehicle starts at a point on the positive z-axis (for $c_4 > 0$). One quarter target orbit period later at $t = T/4$, the vehicle is at the origin $z = 0$ [see Eq. (8.18a)]. Another quarter period later at $t = T/2$ it is on the negative z-axis, and at $t = 3T/4$ it is back at the origin traveling in the positive z-direction. At $t = T$ the vehicle is back to its original point on the positive z-axis. This oscillation along the z-axis is shown in the inertial frame in Fig. 8.5.

As can be seen from Fig. 8.5 the inertial orbit of the vehicle is a circular orbit of the *same period* as the target ($x = \delta r = 0$), but slightly inclined to the target orbit. From the figure the reader can observe that $\tan i$ (which is essentially equal to i in radians because the angle is small) is equal to the maximum value of z divided by r^*. For LEO a maximum z of 100 km then represents an inclination of approximately $1°$.

Now that the general solution for the unthrusted out-of-plane motion has been obtained and interpreted, let us address the general unthrusted in-plane motion, described by the coupled equations:

$$\ddot{x} = 3n^2 x + 2n\dot{y} \qquad (8.15a)$$

$$\ddot{y} = -2n\dot{x} \qquad (8.15b)$$

The general solution is easily obtained by observing that the second equation involves only derivatives, so that it can be integrated directly to yield:

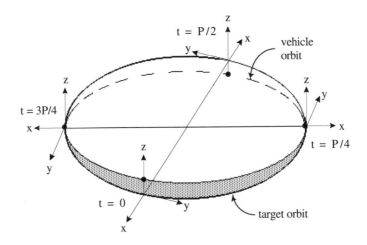

Fig. 8.5 Out-of-plane Oscillation in an Inertial Frame

$$\dot{y} = -2nx + k_1 \tag{8.19}$$

where k_1 is a constant that, evaluated in terms of initial conditions is simply

$$k_1 = \dot{y}_o + 2nx_o \tag{8.20}$$

Substituting $\dot{y}$ from (8.19) into Eq. 8.15a) yields the uncoupled equation:

$$\ddot{x} + n^2 x = 2nk_1 \tag{8.21}$$

which represents an inhomogeneous undamped linear oscillator with a constant forcing term. The x (radial) motion will then be an oscillation of period T about a value $x = 2k_1/n$. When the two arbitrary constants in the solution are evaluated in terms of initial conditions, the result is:

$$x(t) = (4 - 3c)x_o + \frac{s}{n}\dot{x}_o + \frac{2}{n}(1-c)\dot{y}_o \tag{8.22}$$

where $s \equiv \sin nt$ and $c \equiv \cos nt$.

The solution for $y(t)$ from (8.19) will then have two parts: a periodic component due to the oscillatory x-motion and a term of the form $k_1 t + k_2$. The term $k_1 t$ is called *secular* and represents a steady drift along the y-axis whose direction is determined by the algebraic sign of k_1, which is determined by initial conditions according to Eq. (8.20). The solution for the y motion is:

Linear Orbit Theory

$$y(t) = 6(s - nt)x_o + y_o - \frac{2}{n}(1-c)\dot{x}_o + \frac{4s - 3nt}{n}\dot{y}_o \qquad (8.23)$$

An interpretation for these solutions just described is that the oscillation in x ($=\delta r$) representing a varying differential radius describes a neighboring *elliptic* orbit. The drift due to a nonzero value of k_1 describes a neighboring orbit of period slightly different from the target period T, as in the special case of a neighboring circular orbit considered previously. In the special case that the constant k_1 is 0, the periodic x and y motion (of the same period) represents a neighboring elliptic orbit with period *equal to* the target period T.

The combined effects of relative motion in all components of the CW frame then represents the general case of a neighboring orbit which is *elliptic, inclined, and of different period than the target orbit*.

The general solution can be conveniently expressed in terms of the state vector $\delta s(t)$ of the vehicle relative to the target state:

$$\delta s^T(t) = [\ \delta r^T(t)\ \ \delta v^T(t)\ ] = [x\ y\ z\ \dot{x}\ \dot{y}\ \dot{z}\ ] \qquad (8.24)$$

by means of its 6×6 *state transition matrix* $\Phi(t)$ for which:

$$\delta s(t) = \Phi(t)\, \delta s(0) \qquad (8.25)$$

where $\Phi(0) = I_6$, the 6×6 identity matrix. The state transition matrix for the CW equations is obtained by assembling the coefficients of the initial conditions in Eqs. (8.22), (8.23), and (8.18) to yield:

$$\Phi(t) = \begin{bmatrix} 4-3c & 0 & 0 & \frac{s}{n} & \frac{2}{n}(1-c) & 0 \\ 6(s-nt) & 1 & 0 & -\frac{2}{n}(1-c) & \frac{4s-3nt}{n} & 0 \\ 0 & 0 & c & 0 & 0 & \frac{s}{n} \\ 3ns & 0 & 0 & c & 2s & 0 \\ -6n(1-c) & 0 & 0 & -2s & 4c-3 & 0 \\ 0 & 0 & -ns & 0 & 0 & c \end{bmatrix} \qquad (8.26)$$

To perform numerical calculations, it is often convenient to use canonical units, for which $n = 1$ and the reference period $T = 2\pi$ canonical time units.

8.5 Linear Impulsive Rendezvous

A major advantage of the linearized, CW frame is that the standard orbital calculations discussed in previous chapters are greatly simplified. One simply evaluates the state transition matrix $\Phi(t)$ and solves for the state vector $\delta s(t)$. If one partitions the 6×6 transition matrix of Eq. (8.26) into four 3×3 partitions:

$$\Phi(t) = \begin{bmatrix} M(t) & N(t) \\ S(t) & T(t) \end{bmatrix} \qquad (8.27)$$

the relative position vector $\delta \mathbf{r}(t)$ is given by:

$$\delta \mathbf{r}(t) = M(t) \delta \mathbf{r}(0) + N(t) \delta \mathbf{v}(0) \qquad (8.28)$$

where $\delta \mathbf{r}(0)$ and $\delta \mathbf{v}(0)$ represent the six scalar constants that define the orbit relative to the origin of the CW frame. These six constants are equivalent to variations in the six orbital elements discussed in Chap. 3.

Note that Eq. (8.28) represents the linear analog of Kepler's equation, because it tells one the position $\delta \mathbf{r}$ at time t in terms of the state at the initial time $t = 0$.

The corresponding relative velocity $\delta \mathbf{v}(t)$ is given in terms of Eq. (8.27) as:

$$\delta \mathbf{v}(t) = S(t) \delta \mathbf{r}(0) + T(t) \delta \mathbf{v}(0) \qquad (8.29)$$

The orbital boundary value problem, called Lambert's problem in the inverse-square gravitational field and dealt with in Chap. 4, is also greatly simplified. It is instructive to combine the discussion of the boundary value problem with the solution for the impulsive velocity changes necessary to perform a rendezvous. The linear analog of Lambert's problem is also represented by Eq. (8.28) if $\delta \mathbf{r}(t)$ and $\delta \mathbf{r}(0)$ are specified and the unknown is the required initial velocity $\delta \mathbf{v}(0)$.

For simplicity, let us assume the desired final state for the rendezvous is the origin of the CW frame. This simplification can easily be generalized, but if the origin of the CW is placed in a space station, for example, what we are describing is a vehicle performing a rendezvous with the space station. If we let specified final time be denoted by t_f, the desired final relative position is $\delta \mathbf{r}(t_f) = 0$ and the arbitrary initial relative position is simply $\delta \mathbf{r}(0)$. The boundary value problem is then the following: given the initial position $\delta \mathbf{r}(0)$, the desired final position $\delta \mathbf{r}(t_f) = 0$, and the specified transfer time t_f, determine the orbit that satisfies these conditions, including the required initial velocity vector.

Linear Orbit Theory

From Eq. (8.28) one has:

$$\delta \mathbf{r}(t_f) = 0 = M(t_f)\delta\mathbf{r}(0) + N(t_f)\delta\mathbf{v}(0) \tag{8.30}$$

The necessary velocity vector at the initial time is obtained by solving Eq. (8.30) for $\delta\mathbf{v}(0)$ by inverting the partition $N(t_f)$:

$$\delta\mathbf{v}(0) = -N(t_f)^{-1}M(t_f)\delta\mathbf{r}(0) \tag{8.31}$$

and the orbit that solves the boundary value problem is given by substituting into Eq. (8.28):

$$\delta\mathbf{r}(t) = [M(t) - N(t)N(t_f)^{-1}M(t_f)]\delta\mathbf{r}(0) \tag{8.32}$$

which does go to 0 as $t \to t_f$ as it should.

The assumption has been tacitly made that the partition $N(t)$ is invertible for any value of t_f. This is true "almost everywhere," because there exists a countable infinity of values of t_f for which the partition N is singular [8.5]. These values include $t_f = k\pi$, for integer k, and other widely isolated values such as 2.8135π and 4.8906π.

The initial relative velocity vector to intercept the origin is given by Eq. (8.31). In general, an impulsive thrust will be necessary to achieve this velocity, and a better notation for the velocity vector on the left-hand side of Eq. (8.31) would be $\delta\mathbf{v}^+(0)$, which implies that this is the velocity vector after the initial thrust impulse. If one lets the velocity prior to the thrust impulse be $\delta\mathbf{v}^-(0)$, then

$$\delta\mathbf{v}^+(0) = \delta\mathbf{v}^-(0) + \Delta\mathbf{v}_o \tag{8.33}$$

where $\Delta\mathbf{v}_o$ is the velocity change provided by the initial thrust impulse. By combining Eq. (8.33) with Eq. (8.31) one obtains:

$$\Delta\mathbf{v}_o = -N(t_f)^{-1}M(t_f)\delta\mathbf{r}(0) - \delta\mathbf{v}^-(0) \tag{8.34}$$

as the required initial velocity change in terms of the state prior to the impulse.

Equation (8.27) for the relative velocity at the final time can be written as:

$$\delta\mathbf{v}^-(t_f) = S(t_f)\delta\mathbf{r}(0) + T(t_f)\delta\mathbf{v}^+(0) \tag{8.35}$$

where the minus superscript at the final time denotes conditions prior to the thrust impulse at the final time. By combining Eqs. (8.31) and (8.35) one obtains:

$$\delta\mathbf{v}^-(t_f) = [S(t_f) - T(t_f)N(t_f)^{-1}M(t_f)]\delta\mathbf{r}(0) \tag{8.36}$$

The final velocity change must satisfy:

$$\delta v^+(t_f) = 0 = \delta v^-(t_f) + \Delta v_f. \tag{8.37}$$

Combining Eqs. (8.36) and (8.37) yields the final velocity change:

$$\Delta v_f = [T(t_f)N(t_f)^{-1}M(t_f) - S(t_f)]\delta r(0) \tag{8.38}$$

Equations (8.32) and (8.36) provide closed form expressions for the two velocity changes required to perform the rendezvous. Physically, the initial velocity change places the vehicle on a trajectory which will intercept the origin at time t_f. The final velocity change cancels the arrival velocity of the vehicle at the origin and completes the rendezvous.

Example 8.1

As an example of impulsive rendezvous, consider a Hohmann transfer in the CW frame. The vehicle is initially in a circular orbit, as illustrated in Fig. 8.4. In units such that n is equal to unity, the Hohmann transfer time is equal to $t_f = \pi$ time units. Thus $s = 0$ and $c = -1$ in the state transition matrix of Eq. (8.26). Also, if one assumes a coplanar problem, $z \equiv 0$ and one can use only the upper left 2×2 submatrices of the partitions M, N, S, and T of Eq. (8.27). In particular,

$$M(t_f) = \begin{bmatrix} 7 & 0 \\ -6\pi & 1 \end{bmatrix} \; ; \; N(t_f) = \begin{bmatrix} 0 & 4 \\ -4 & -3\pi \end{bmatrix} \tag{8.39}$$

The initial orbit is a circular orbit represented by a specified differential radius x_o, along with $\dot{x}_o = 0$ and $\dot{y}_o = -1.5 x_o$, as given by Eq. (8.17). The initial value of y_o for a Hohmann transfer to the origin of the CW frame has to be determined. This value plays the same role as the lead/lag angle β_{12} or β_{21} in Sec. 6.7, indicating where, relative to the target, the first impulse should be applied.

Solving Eq. (8.31) for the value of $\delta v(0)$ that has only a y component, indicating the transfer orbit is tangent to the initial circular orbit, yields $y_o = 0.75\pi x_o$ as the required value. The resulting transfer orbit, given by Eq. (8.32), arrives at the origin with only a y-component of velocity. This is shown in Fig. 8.6. The transfer path is obtained by solving Eq. (8.32) and the two required velocity changes are given by Eqs. (8.34) and (8.38).

Linear Orbit Theory

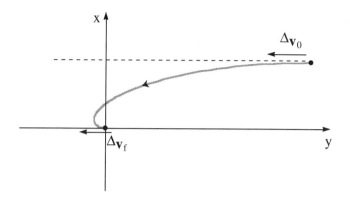

Fig. 8.6 A Hohmann Transfer in the CW Frame

References

8.1 Clohessy, W. H., and Wiltshire, R. S., "Terminal Guidance Systems for Satellite Rendezvous," *Journal of the Aerospace Sciences*, Sept. 1960, pp. 653–658, and 674.

8.2 Hill, G. W., "Researches in the Lunar Theory," *American Journal of Mathematics*, **1**, 1878, pp. 5–26.

8.3 Kaplan, M. H., *Modern Spacecraft Dynamics and Control*, John Wiley and Sons, 1976.

8.4 Gobetz, F. W., "Optimal Variable-Thrust Transfer of a Power-Limited Rocket Between Neighboring Circular Orbits," *AIAA Journal*, **2**, 2, 1964.

8.5 Prussing, J. E. and Clifton, R. S., "Optimal Multiple-Impulse Satellite Avoidance Maneuvers," Preprint AAS 87-543, AAS/AIAA Astrodynamics Specialist Conference, Kalispell MT, Aug. 1987.

Problems

8.1 Verify the singularities of the N partition of the state transition matrix mentioned following Eq. (8.32).

8.2 Generalize the expressions (8.34) and (8.38) for the velocity changes for an *arbitrary* final relative state $\delta\mathbf{r}(t_f)$ and $\delta\mathbf{v}^+(t_f)$.

8.3 Briefly explain why the expression for $\Delta\mathbf{v}_f$ in Eq. (8.38) is not a function of the initial velocity $\delta\mathbf{v}^-(0)$, whereas $\Delta\mathbf{v}_o$ in (8.32) is.

8.4 Determine the required velocity changes to rendezvous with the origin for $\delta\mathbf{r}(0)^T = [\,1\ 1\ 1\,]$, $\delta\mathbf{v}^-(0)^T = [\,0\ 0\ 1\,]$, and $t_f = \pi/2$. Use units for which $n = 1$.

8.5 Determine the required velocity changes to rendezvous with the origin for $\delta\mathbf{r}(0)^T = [\,0\ 1\ 0\,]$, $\delta\mathbf{v}^-(0) = \mathbf{0}$, and $t_f = 2\pi$. Is it possible to solve for the velocity changes even though the N partition is singular? If so, briefly discuss how this can be possible in light of Eqs. (8.34) and (8.38), which require $N^{-1}(t_f)$.

8.6* Derive the expression for the gravity gradient matrix for an inverse-square gravitational field shown in Eq. (8.9).

8.7* For the Hohmann transfer considered in Ex. 8.1, calculate the required velocity changes.

8.8* Determine an analytical expression for the matrix $N^{-1}(t)$ utilized in Eq. (8.31) and thereafter.

9
Perturbation

9.1 Introduction

The discussion of satellite motion in the preceding chapters has assumed a two-body system in which the central body is a sphere having uniform or radially symmetric mass distribution that acts gravitationally, as shown in Sec. 1.3, as a point mass. This is an approximation based on ignoring certain effects that need to be included in many practical applications. A satellite in earth orbit experiences small but significant perturbations (accelerations) due to the lack of spherical symmetry of the earth, the attraction of the moon and sun for the satellite, and, if the satellite is in low orbit, due to atmospheric drag. Any of these perturbations alone is sufficient to cause predictions of the position of a satellite based on a Keplerian or two-body unperturbed model to be in significant error in a brief time.

The equations that describe the variation in time of the orbit elements due to these perturbations, which are usually called the *perturbation equations*, will be derived here in their simplest form, employing the conventional orbit elements and containing explicitly the components of the disturbing forces. There are many ways of deriving these perturbation equations. The methods most often found in the literature [9.1, 9.2] apply powerful and mathematically elegant tools of classical mechanics that are unnecessary. It is possible to derive the same result using first principles (e.g., Newton's Second Law of motion) and thereby retain much greater physical insight. This is the approach of Burns [9.3] that will be followed here.

9.2 The Perturbation Equations

The problem then is to determine the time rates of change of the orbital elements $(a, e, i, \Omega, \omega, M)$ in the presence of an external force. We will assume that the magnitude of the disturbing force is much smaller than the magnitude of the attraction of the satellite for the primary; this is true for all of the perturbations mentioned in the preceding section. It will be convenient to express the perturbation as an acceleration, that is, as a force per unit mass. This disturbing acceleration may then be resolved into components as

$$\mathbf{F} = R\hat{\mathbf{e}}_R + T\hat{\mathbf{e}}_T + N\hat{\mathbf{e}}_N \qquad (9.1)$$

where the unit vectors comprise a rotating basis whose origin is fixed to the satellite, as shown in Fig. 9.1. The unit vector $\hat{\mathbf{e}}_R$ points radially outward; $\hat{\mathbf{e}}_T$ is normal to $\hat{\mathbf{e}}_R$, lies in the orbit plane and is positive in the direction of motion; $\hat{\mathbf{e}}_N$ is then defined as $\hat{\mathbf{e}}_R \times \hat{\mathbf{e}}_T$ and is normal to the orbit plane.

In the general perturbed problem there are no longer "constants of the motion" as there were in the Kepler problem. There are special cases; for example, if the disturbing force is purely radial the orbit experiences no torque and so its angular momentum is conserved, but its mechanical energy will not be conserved. Quantities defined for the unperturbed problem such as the mechanical energy (per unit mass),

$$\varepsilon = \frac{v^2}{2} - \frac{\mu}{r} = -\frac{\mu}{2a} \qquad (9.2,\ 2.18,\ 2.22)$$

or angular momentum (per unit mass)

$$\mathbf{h} = \mathbf{r} \times \mathbf{v}, \quad \text{where} \quad h = [\mu a(1-e^2)]^{1/2} \qquad (9.3,\ 1.35)$$

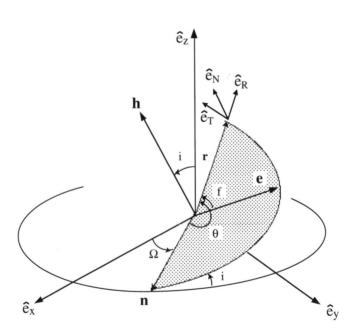

Fig. 9.1 Orbital Elements and Unit Vector Bases

Perturbation

still describe the orbit but are now functions of the *instantaneous* values of the orbital elements. The orbit semimajor axis and eccentricity may be expressed in terms of ε and h. The time rates of change of a and e will then be functions of the time rates of change of ε and h.

From Eq. (9.2) we have

$$a = -\frac{\mu}{2\varepsilon} \tag{9.4}$$

so that

$$\dot{a} = \frac{\mu}{2\varepsilon^2}\dot{\varepsilon} = 2a^2\mu^{-1}\dot{\varepsilon} \tag{9.5}$$

We see immediately that any perturbation that causes the dissipation of mechanical energy from the system, for example atmospheric drag, will cause the semimajor axis to contract.

From Eq. (9.3)

$$e^2 = 1 - \frac{h^2}{\mu a} \tag{9.6}$$

Substituting from Eq. (9.4) for a yields,

$$e = (1 + 2h^2\varepsilon\mu^{-2})^{\frac{1}{2}} \tag{9.7}$$

so that

$$\dot{e} = \frac{1}{2e}(e^2 - 1)\left[\frac{2\dot{h}}{h} + \frac{\dot{\varepsilon}}{\varepsilon}\right] \tag{9.8}$$

Now from the principle of work and energy we have

$$d\varepsilon = \mathbf{F} \cdot d\mathbf{r} \tag{9.9}$$

so that

$$\dot{\varepsilon} = \mathbf{F} \cdot \dot{\mathbf{r}} = \mathbf{F} \cdot (\dot{r}\hat{\mathbf{e}}_R + r\dot{\theta}\hat{\mathbf{e}}_T) \tag{9.10}$$

or, substituting from Eq. (9.1),

$$\dot{\varepsilon} = \dot{r}R + r\dot{\theta}T \tag{9.11}$$

From Newton's second law (for rotational motion)

$$\dot{\mathbf{h}} = \mathbf{r} \times \mathbf{F} = rT\hat{\mathbf{e}}_N - rN\hat{\mathbf{e}}_T \tag{9.12}$$

but the $\hat{\mathbf{e}}_T$ component of $\dot{\mathbf{h}}$ is always normal to $\mathbf{h}$ itself, because $\mathbf{h}$ is by definition purely in the $\hat{\mathbf{e}}_N$ direction. Therefore only the $\hat{\mathbf{e}}_N$ component of $\dot{\mathbf{h}}$ changes the magnitude of $\mathbf{h}$, or

$$\dot{h} = rT \qquad (9.13)$$

From the definition of angular momentum we have, assuming ω small,

$$\dot{\theta} = \dot{f} = h/r^2 \qquad (9.14)$$

and from the radial distance equation

$$r = \frac{h^2/\mu}{1 + e\cos f} \qquad (9.15, 1.32)$$

Therefore

$$\dot{r} = \frac{r^2 e \sin f \, \dot{f}}{h^2/\mu} = \frac{\mu e \sin f}{h} \qquad (9.16)$$

using Eqs. (9.14) and (9.3). Substituting these results for $\dot{r}$ and $\dot{\theta}$ in Eqs. (9.11) and (9.5) yields

$$\dot{a} = 2a^2\mu^{-1} \left[R \frac{\mu e \sin f}{h} + T \frac{h}{r} \right] \qquad (9.17)$$

or, since $\mu/h = [\mu/a(1-e^2)]^{1/2}$,

$$\dot{a} = 2a^{3/2}\mu^{-1/2}(1-e^2)^{-1/2} [R e \sin f + T(1 + e \cos f)] \qquad (9.18)$$

Similarly, since $\dot{\varepsilon} = \dot{a}\mu/2a^2$ from Eq. (9.5) Eq. (9.8) may be written as

$$\dot{e} = \frac{1}{2}e^{-1}(e^2 - 1) \cdot$$

$$\left\{ \frac{2rT}{h} + \frac{1}{\varepsilon} \frac{\mu^{1/2}}{a^{1/2}(1-e^2)^{1/2}} [R e \sin f + T(1 + e \cos f)] \right\} \qquad (9.19)$$

or

$$\dot{e} = \frac{e^2 - 1}{2e} \left\{ \frac{-2a^{1/2}}{\mu^{1/2}(1-e^2)^{1/2}} R e \sin f \right. \qquad (9.20)$$

$$\left. + \left[\frac{2a(1 - e\cos E)}{(\mu a(1-e^2))^{1/2}} - \frac{2a^{1/2}}{\mu^{1/2}(1-e^2)^{1/2}} (1 + e \cos f) \right] T \right\}$$

Because

$$r = a(1 - e\cos E) \qquad (9.21, 2.12)$$

Eq. (9.20) simplifies to

$$\dot{e} = [a(1-e^2)\mu^{-1}]^{1/2} [R \sin f + T(\cos f + \cos E)] \qquad (9.22)$$

Perturbation

The *variational equations* for $\dot{a}$ and $\dot{e}$ are perhaps the simplest to obtain, but it is also straightforward to find di/dt. From Fig. 9.1 it is clear that

$$\cos i = \frac{(\mathbf{h} \cdot \hat{\mathbf{e}}_z)}{h} = \frac{h_z}{h} \tag{9.23}$$

Therefore,

$$\frac{di}{dt} = \frac{1}{-\sin i} \frac{h \dot{h}_z - \dot{h} h_z}{h^2} \tag{9.24}$$

or,

$$\frac{di}{dt} = \frac{1}{(h^2 - h_z^2)^{1/2}} \frac{h \dot{h}_z - \dot{h} h_z}{h} = \left(\frac{\dot{h}}{h} - \frac{\dot{h}_z}{h_z} \right) \frac{1}{\tan i} \tag{9.25}$$

But,

$$\dot{h}_z = \frac{d\mathbf{h}}{dt} \cdot \hat{\mathbf{e}}_z = (rT\hat{\mathbf{e}}_N - rN\hat{\mathbf{e}}_T) \cdot \hat{\mathbf{e}}_z \tag{9.26}$$

$$= rT \cos i - rN \cos \theta \sin i$$

and

$$h_z = \mathbf{h} \cdot \hat{\mathbf{e}}_z = (\mu a (1 - e^2))^{1/2} \cos i \tag{9.27}$$

Substituting from Eqs. (9.27) and (9.26) into Eq. (9.25) yields

$$\frac{di}{dt} = \frac{[a\mu^{-1}(1 - e^2)]^{1/2} N \cos \theta}{(1 + e \cos f)} \tag{9.28}$$

From Fig. 9.1 it is clear by inspection that

$$h_x = \mathbf{h} \cdot \hat{\mathbf{e}}_x = h \sin i \sin \Omega \tag{9.29a}$$

and

$$h_y = \mathbf{h} \cdot \hat{\mathbf{e}}_y = -h \sin i \cos \Omega \tag{9.29b}$$

so that

$$\tan \Omega = h_x / (-h_y) \tag{9.30}$$

or

$$\frac{1}{\cos^2 \Omega} \dot{\Omega} = \frac{h_x \dot{h}_y - h_y \dot{h}_x}{h_y^2} \tag{9.31}$$

But

$$\cos^2 \Omega = \frac{1}{1 + \tan^2 \Omega} = \frac{h_y^2}{h_x^2 + h_y^2} \quad (9.32)$$

Therefore, from Eq. (9.31),

$$\dot{\Omega} = \frac{h_x \dot{h}_y - h_y \dot{h}_x}{h_x^2 + h_y^2} \quad (9.33)$$

From Eq. (9.12), $\dot{h}_x$ and $\dot{h}_y$ may be found as

$$\dot{h}_x = \hat{e}_x \cdot (rT\hat{e}_N - rN\hat{e}_T) \quad (9.34a)$$

$$\dot{h}_y = \hat{e}_y \cdot (rT\hat{e}_N - rN\hat{e}_T) \quad (9.34b)$$

In order to take the indicated dot products the transformation from rotating to fixed basis vectors is required. Observing from Fig. 9.1 that the angles Ω, i, and θ represent Euler angles for this transformation, we have

$$\begin{bmatrix} \hat{e}_x \\ \hat{e}_y \\ \hat{e}_z \end{bmatrix} = \begin{bmatrix} c\theta c\Omega - ci\ s\Omega s\theta & -s\theta c\Omega - ci\ s\Omega c\theta & si\ s\Omega \\ c\theta s\Omega + ci\ c\Omega s\theta & -s\theta s\Omega + ci\ c\Omega c\theta & -si\ c\Omega \\ si\ s\theta & si\ c\theta & ci \end{bmatrix} \begin{bmatrix} \hat{e}_R \\ \hat{e}_T \\ \hat{e}_N \end{bmatrix} \quad (9.35)$$

Using Eq. (9.35) to solve for $\dot{h}_x$ and $\dot{h}_y$ and employing Eq. (9.27) for h_z yields

$$\dot{\Omega} = \frac{[a\mu^{-1}(1-e^2)]^{1/2} N \sin\theta}{\sin i\ (1 + e\cos f)} \quad (9.36)$$

The remaining two perturbation equations for $\dot{\omega}$ and $\dot{M}$ are more difficult to derive because these two orbit elements are not explicit functions of ε or $\mathbf{h}$. However, we may write

$$h^2 = \mu a(1 - e^2) = \mu r(1 + e\cos f)$$

$$= \mu r[1 + e\cos(\theta - \omega)] \quad (9.37a)$$

$$= \mu r[1 + (1 + 2\varepsilon h^2 \mu^{-2})^{1/2} \cos(\theta - \omega)]$$

Taking the time derivative of Eq. (9.37a) yields

$$2h\dot{h} = \mu r\ (-(1 + 2\varepsilon h^2 \mu^{-2})^{1/2}\ (\dot{\theta} - \dot{\omega}) \sin(\theta - \omega)$$

$$+ \cos(\theta - \omega) \left[\frac{1}{2}(1 + 2\varepsilon h^2 \mu^{-2})^{-1/2} (2\dot{\varepsilon}h^2\mu^{-2} + 4\varepsilon h\dot{h}\mu^{-2}) \right]$$

Perturbation 161

$$+ \mu\dot{r}[1 + (1 + 2\varepsilon h^2\mu^{-2})^{\frac{1}{2}} \cos(\theta - \omega)]). \tag{9.37b}$$

In the absence of any perturbing accelerations, $\dot{h}$ and $\dot{\varepsilon}$ are 0; therefore, it must be true that

$$0 = \mu r \left[-(1 + 2\varepsilon h^2\mu^{-2})^{\frac{1}{2}} (\dot{\theta} - \dot{\omega}) \sin(\theta - \omega) \right]$$

$$+ \mu\dot{r} \left[1 + (1 + 2\varepsilon h^2\mu^{-2})^{-\frac{1}{2}} \cos(\theta - \omega) \right] \tag{9.37c}$$

which can be simply written as

$$0 = \mu r(-e\dot{f} \sin f) + \mu\dot{r}(1 + e \cos f) \tag{9.37d}$$

Equation (9.37c) is satisfied at any time, even when perturbations are acting, by the *instantaneous* value or r, e, and f (or ε, h, θ, and ω). The terms on the right-hand side of Eq. (9.37c) may then be removed from the right-hand side of (9.37d) to yield:

$$2h\dot{h} = \mu r [-e(\dot{\theta} - \dot{\omega})_{pert} \sin(\theta - \omega)$$

$$+ \cos(\theta - \omega) \cdot \frac{1}{2} e^{-1} \mu^{-2} (2h^2\dot{\varepsilon} + 4\varepsilon h\dot{h})] \tag{9.37e}$$

where $(\dot{\theta} - \dot{\omega})_{pert}$ indicates that part of $(\dot{\theta} - \dot{\omega})$ caused exclusively by the perturbing forces. Equation (9.37e) may be rewritten as

$$\dot{\omega} = (\dot{\theta})_{pert} - \frac{h^2\dot{\varepsilon} \cot(\theta - \omega)}{\mu^2 e^2}$$

$$+ \frac{2h\dot{h}}{e\mu \sin(\theta - \omega)} \left[\frac{1}{r} - \frac{\varepsilon \cos(\theta - \omega)}{e\mu} \right] \tag{9.38}$$

Change in θ due exclusively to the perturbation is illustrated in Fig. 9.2. If the longitude of the ascending node, Ω, changes then θ must change as well, since θ is measured from the ascending node. By inspection, change in Ω is related to change in θ by $d\theta = -d\Omega \cos i$, so that

$$(\dot{\theta})_{pert} = -\dot{\Omega} \cos i \tag{9.39}$$

Then, substituting this result as well as $\dot{\varepsilon}$ from Eq. (9.17) and $\dot{h}$ from Eq. (9.13) into Eq. (9.38) yields

$$\dot{\omega} = -\dot{\Omega} \cos i + e^{-1}[\mu^{-1} a(1 - e^2)]^{\frac{1}{2}} \cdot$$

$$\left[-R \cos f + \frac{T \sin f (2 + e \cos f)}{1 + e \cos f} \right] \tag{9.40}$$

The mean anomaly has previously (Sec. 2.2) been defined as

$$M = n(t - \tau) = E - e \sin E \qquad (9.41, 2.7)$$

where E is the eccentric anomaly and τ is the time of periapse passage. If we define an auxiliary variable $\chi = n\tau$, then Eq. (9.41) becomes $M = nt - \chi$ so that

$$\dot{M} = n + \dot{n}t - \dot{\chi} = \dot{E} - e\dot{E} \cos E - \dot{e} \sin E \qquad (9.42)$$

or

$$\dot{\chi} = n + \dot{n}t - \dot{E}(1 - e \cos E) + \dot{e} \sin E \qquad (9.43)$$

or

$$\dot{\chi} = n - \frac{3n\dot{a}t}{2a} - \dot{E}(1 - e \cos E) + \dot{e} \sin E \qquad (9.44)$$

since $\dot{n} = d/dt\, (\mu/a^3)^{1/2} = -3n\dot{a}/2a$. The derivatives $\dot{a}$ and $\dot{e}$ are known from Eq. (9.18) and (9.22). $\dot{E}$ may be found by taking time derivatives of both sides of Eq. (2.12) to be

$$\dot{E} = \frac{\dot{r} - \dot{a}(1 - e \cos E) + a\dot{e} \cos E}{ae \sin E} \qquad (9.45)$$

where $\dot{r}$ is given by Eq. (9.16). Equation (9.44) becomes the sixth (and last) perturbation equation. When the orbital element M needs to be known it may be determined from $M = nt - \chi$ where χ is found by integration of (9.44) and n is determined by the initial value of a; that is, the value a has at the start of the integration of the system of six perturbation equations.

Unfortunately, integration of Eq. (9.44) is difficult because of the presence of the term explicitly containing t. This implies that χ will get very large, and $\dot{\chi}$ even larger, if the integration continues over large spans of time. The solution to this difficulty is to define a new variable $\bar{\chi}$ such that

$$M = nt - \chi = \int n\, dt - \bar{\chi} = E - e \sin E \qquad (9.46)$$

Then

$$\dot{\bar{\chi}} = n - \dot{E}(1 - e \cos E) + \dot{e} \sin E \qquad (9.47)$$

so that the explicit dependence on t is removed. Note that n appearing in Eq. (9.47) now represents the instantaneous value, that is, $n = (\mu/a^3)^{1/2}$. The "cost" of this method is that an auxiliary equation must be added to the original system of six perturbation equations, because to evaluate M at any time from Eq. (9.46) it is necessary to evaluate $\int n\, dt$. This can be done by

Perturbation

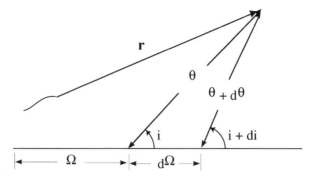

Fig. 9.2 Change in θ Due to a Change in Ω

simply integrating a seventh equation:

$$\dot{\xi} = n = (\mu/a^3)^{1/2} \tag{9.48}$$

so that ξ will represent $\int n \, dt$.

By substituting for $\dot{E}$ and $\dot{e}$ in Eq. (9.47) and employing the identities

$$\cos E = \frac{e + \cos f}{1 + e \cos f} \tag{9.49, 2.9}$$

$$\sin E = \frac{(1 - e^2)^{1/2} \sin f}{1 + e \cos f}$$

it is possible to obtain $\dot{\chi}$ in the form:

$$\dot{\chi} = \frac{a^{1/2} \mu^{-1/2} (1 - e^2)}{e (1 + e \cos f)} [R (2e - \cos f - e \cos^2 f)$$

$$+ T \sin f (2 + e \cos f)] \tag{9.50}$$

To summarize, the six equations for the time rates of change of the orbital elements, which are usually referred to as Gauss' form of the perturbation equations, are given by Eqs. (9.18), (9.22), (9.28), (9.36), (9.40), and (9.50). Because these equations possess singularities for $e = 0$ and $\sin i = 0$, an analogous set of equations that are free of singularities can be derived using the equinoctial elements discussed in Prob. 3.7.

9.3 Effect of Atmospheric Drag

With certain assumptions, which are reasonable for satellites in low earth orbit, the effect of drag can be simply obtained. The drag force acts in a direction opposite to the velocity so that, if the orbit is nearly circular,

$$N = 0, \quad R = 0, \quad T = -\tfrac{1}{2}\rho C_D A v^2 / m \tag{9.51}$$

where C_D is the *drag coefficient*, ρ is the atmospheric density, and A is the cross-sectional area of the satellite presented to the flow. The product $C_D A / m$ is called the *ballistic coefficient* of the satellite. Then from Eq. (9.17)

$$\dot{a} = 2\mu^{-1/2} a^{3/2} T(1 + e \cos f) = -2\mu^{-1/2} a^{3/2} [\rho C_D A v^2 / 2m]$$

$$= -(\mu a)^{1/2} \rho C_D A / m < 0 \tag{9.52}$$

where the fact that $v = (\mu/a)^{1/2}$ for a nearly circular orbit has been used. Then

$$\frac{da}{a^{1/2}} = \frac{-\mu^{1/2} \rho C_D A}{m} dt \tag{9.53}$$

so that

$$a_{final}^{1/2} - a_{initial}^{1/2} = \frac{-\mu^{1/2} \rho C_D A (t_f - t_i)}{2m} \tag{9.54}$$

This result for Δa, however, is only valid when Δa is small; otherwise, it is not appropriate to consider atmospheric density constant in the integration of Eq. (9.53).

Because there is no component of drag normal to the orbit we see from Eqs. (9.28) and (9.36) that Ω and i are unaffected by drag, that is, *the orientation of the orbit plane is not changed by drag.*

9.4 Effect of Earth Oblateness

The gravitational field of the earth is not spherically symmetric, as we have assumed throughout most of the book, however the nonuniformities produce accelerations that are very small in comparison with the central-body attraction. As such their effect on the orbit may be treated as a small perturbation to the Keplerian motion. There are both longitude- and latitude-dependent variations in the gravitational field, but the principal nonspherical variation is the "bulging" of the earth at the equator due to the earth's rotation. The gravitational potential of the earth, neglecting terms that have a longitudinal variation, is

$$U = \frac{\mu}{r} \left[1 - \sum_{k=2}^{\infty} \left(\frac{R}{r}\right)^k J_k P_k(\cos\phi) \right] \tag{9.55}$$

Perturbation

where R is the radius of the earth, the J_k are constant coefficients, the P_k are *Legendre polynomial functions* (of order k), and ϕ is the *colatitude*, illustrated in Fig. 9.3. This potential function U is the negative of the potential energy V discussed in Chap. 1.

The first term in Eq. (9.55) represents the familiar spherically symmetric central body attraction. The other terms represent the disturbing potential and of these the first, the *oblateness* term, is by far the principal one, J_2 being order 10^3 times as large as any of the other gravity coefficients. From Eq. (9.55) this disturbing potential is

$$U_{J_2} = -\frac{\mu}{r}\left(\frac{R}{r}\right)^2 J_2 P_2(\cos\phi)$$

$$= \frac{-\mu R^2}{r^3} J_2 \left[\frac{1}{2}(2 - 3\sin^2\phi)\right] \quad (9.56)$$

From Fig. 9.3 it is clear that $\sin^2\phi = 1 - z^2/r^2$, therefore

$$U_{J_2} = \frac{-\mu R^2}{r^3} J_2 \left[\frac{1}{2}\left(\frac{3z^2}{r^2} - 1\right)\right] \quad (9.57)$$

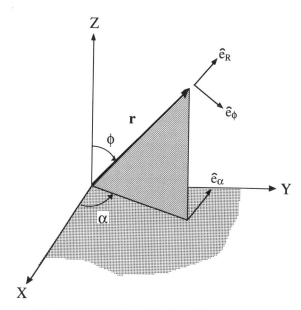

Fig. 9.3 Unit Vector Basis and Colatitude ϕ

a form that will be more convenient for the determination of the disturbing acceleration. This perturbation is the gradient of the potential (9.57), or

$$\mathbf{F}_{obl} = \nabla U_{J_2} = \frac{\partial U}{\partial r}\hat{\mathbf{e}}_R + \frac{\partial U}{\partial z}\hat{\mathbf{e}}_z$$

$$= -\mu J_2 R^2 \left[\frac{3z}{r^5}\hat{\mathbf{e}}_z + \left(\frac{3}{2r^4} - \frac{15z^2}{2r^6}\right)\hat{\mathbf{e}}_R\right] \quad (9.58)$$

But from Eq. (9.35)

$$\hat{\mathbf{e}}_z = \sin i \, \sin\theta \, \hat{\mathbf{e}}_R + \sin i \, \cos\theta \, \hat{\mathbf{e}}_T + \cos i \, \hat{\mathbf{e}}_N \quad (9.59)$$

and since

$$z = r \cos\phi = r \sin i \sin\theta \quad (9.60)$$

the perturbing acceleration becomes,

$$\mathbf{F}_{obl} = \frac{-3\mu J_2 R^2}{r^4}\left[\hat{\mathbf{e}}_R\left(\frac{1}{2} - \frac{3\sin^2 i \, \sin^2\theta}{2}\right) + \hat{\mathbf{e}}_T \sin^2 i \, \sin\theta \cos\theta\right.$$

$$\left. + \hat{\mathbf{e}}_N \sin i \, \sin\theta \cos i\right] \quad (9.61)$$

The principal, secular effects of this disturbance are found in the motion of the ascending node and line of apsides. Substituting for $\mathbf{F}_{obl}$ into Eq. (9.36) yields

$$\dot{\Omega} = \frac{[a\mu^{-1}(1-e^2)]^{1/2}\sin\theta}{\sin i \, (1 + e \cos f)}\left[\frac{-3\mu J_2 R^2}{r^4}\sin i \, \sin\theta \cos i\right] \quad (9.62)$$

or

$$\dot{\Omega} = \frac{-3\mu J_2 R^2}{hp^3}\cos i \, \sin^2\theta \, [1 + e \cos(\theta - \omega)]^3 \quad (9.63)$$

where the semilatus rectum $p = a(1 - e^2)$. Note that $\Omega < 0$ for a posigrade $(0 \le i \le \pi/2)$ orbit, indicating that the line of nodes regresses (moves westward). To determine the average rate of change of Ω, over many orbits, it is convenient to replace time with argument of latitude θ as the independent variable. Then

$$\frac{d\Omega}{d\theta} = \frac{\dot{\Omega}}{\dot{\theta}} \approx \frac{\dot{\Omega}}{h/r^2} = -3J_2(R/p)^2 \cos i \, \sin^2\theta \, [1 + e \cos(\theta - \omega)] \quad (9.64)$$

The average value of $d\Omega/d\theta$ is then obtained as

$$\left(\frac{d\Omega}{d\theta}\right)_{AV} = -3J_2(R/p)^2 \cos i \, \frac{1}{2\pi}\int_0^{2\pi} \sin^2\theta \, [1 + e \cos(\theta - \omega)] \, d\theta \quad (9.65)$$

Perturbation

The required integral may be expressed as

$$\int_0^{2\pi} \sin^2\theta \, (1 + e \cos\theta \cos\omega + e \sin\theta \sin\omega) \, d\theta$$

$$= \int_0^{2\pi} \sin^2\theta \, d\theta + e \cos\omega \int_0^{2\pi} \sin^2\theta \cos\theta \, d\theta$$

$$+ e \sin\omega \int_0^{2\pi} \sin^3\theta \, d\theta$$

$$= \pi + 0 + 0 = \pi \tag{9.66}$$

Therefore,

$$\left(\frac{d\Omega}{d\theta}\right)_{AV} = \frac{-3}{2} J_2 \left(\frac{R}{p}\right)^2 \cos i \tag{9.67}$$

or,

$$\dot{\Omega}_{AV} = \left(\frac{d\Omega}{d\theta}\right)_{AV} \cdot \dot{\theta}_{AV} = \left(\frac{d\Omega}{d\theta}\right)_{AV} \cdot n = \frac{-3n}{2} J_2 \left(\frac{R}{p}\right)^2 \cos i \tag{9.68}$$

where J_2 for earth $= 1082.7 \times 10^{-6}$.

For a typical space shuttle orbit that is nearly circular, has an inclination of 28°, and an altitude of 300 km, Eq. (9.67) yields $\dot{\Omega} = -1.51 \times 10^{-6}$ rad/s $= -7.49°$/day, a very significant perturbation.

Equation (9.40) for the motion of the argument of perigee may be written as

$$\dot{\omega} + \dot{\Omega} \cos i = e^{-1}[a\mu^{-1}(1-e^2)]^{1/2} \cdot$$

$$\left[-R \cos(\theta - \omega) + \frac{T \sin(\theta - \omega)[2 + e \cos(\theta - \omega)]}{1 + e \cos(\theta - \omega)} \right] \tag{9.69}$$

The mean value of the quantity

$$\frac{d\omega}{d\theta} + \frac{d\Omega}{d\theta} \cos i$$

may be found in the same manner as $(d\Omega/d\theta)_{AV}$ was found with the result

$$\left[\frac{d\omega}{d\theta} + \frac{d\Omega}{d\theta} \cos i\right]_{AV} = \frac{3J_2 R^2}{2p^2} \left[1 - \frac{3}{2}\sin^2 i\right] \tag{9.70}$$

Substituting from Eq. (9.67) allows $(d\omega/d\theta)_{AV}$ to be isolated as

$$\left[\frac{d\omega}{d\theta}\right]_{AV} = \frac{3}{2}J_2\left(\frac{R}{p}\right)^2\left[2 - \frac{5}{2}\sin^2 i\right] \qquad (9.71)$$

Comparing this result with Eq. (9.67) shows that $\dot{\omega}$ and $\dot{\Omega}$ are of the same magnitude. From Eq. (9.70), as the orbit plane approaches the equatorial plane:

$$\left[\frac{d}{d\theta}(\Omega + \omega)\right]_{AV} \rightarrow \frac{3}{2}J_2\left(\frac{R}{p}\right)^2 \qquad (9.72)$$

that is, the longitude of perigee $(\Omega + \omega)$ advances by $3\pi J_2 (R/p)^2$ radians per revolution.

References

9.1 Moulton, F. R., *An Introduction to Celestial Mechanics*, Macmillan, New York, 1970

9.2 Plummer, H. C., *An Introductory Treatise on Dynamical Astronomy*, Dover, New York, 1970.

9.3 Burns, J. A.,"Elementary Derivation of the Perturbation Equations of Celestial Mechanics," *Am. J. Physics*, **44**, 10, 944–949, 1976.

Problems

9.1 Derive Eq. (9.50) for the time variation of the auxiliary variable χ following the method suggested in the text.

9.2 Derive Eq. (9.71) for the average variation of ω with θ using the suggestions in the text.

9.3 Show that a consideration of the earth's oblateness for first order (terms involving J_2 only) has no net effect on orbital inclination, eccentricity, and semimajor axis. That is, determine $di/d\theta$, $de/d\theta$, and $da/d\theta$, and average these variations over one orbit.

9.4 Determine the value of the semimajor axis of a *sun-synchronous* orbit with $e = 0$, $i = 97°$.

9.5 Taking into account the rotation of the upper atmosphere with the earth, the drag acceleration on a satellite is

$$d\mathbf{F}_{drag} = -\frac{1}{2}C_D \rho A v_{rel} \mathbf{v}_{rel}$$

$$= R\hat{\mathbf{e}}_R + N\hat{\mathbf{e}}_N + T\hat{\mathbf{e}}_T$$

where C_D is the drag coefficient, ρ is the (constant) atmospheric density, and

$$\mathbf{v}_{rel} = \mathbf{v} - \omega_e \,\hat{\mathbf{e}}_z \times r\hat{\mathbf{e}}_R$$

where ω_e is the rotational angular velocity of the earth.

a) Assuming a circular orbit and neglecting terms of order $(\omega_e/n)^2$ show that:

$$v_{rel} = na\left(1 - \frac{\omega_e}{n}\cos i\right)$$

$$R = 0$$

$$T = -\frac{AC_D}{2}\rho a^2 n^2\left(1 - \frac{\omega_e}{n}\cos i\right)$$

$$N = -\frac{AC_D}{2}\rho a^2 n\, \omega_e\, \sin i\, \cos\theta$$

b) Find the average (over θ) rate of change per unit time of the orbital

1) energy, where $d\varepsilon/dt = \mathbf{v}\cdot d\mathbf{F}$
2) angular momentum (scalar), where $dh/dt = rT$.
3) semimajor axis
4) inclination
5) longitude of the ascending node, Ω

10
Orbit Determination

10.1 Introduction

The subjects of this chapter are orbit determination from observations, sometimes called *initial orbit determination* (IOD), and the improvement of an orbit from observations, sometimes called *differential correction* (DC). The literature on each subject is exhaustive; there are different methods of IOD for different types of orbits, different types of observational data, and different amounts of observational data. Similarly, there are different methods of DC, the principal distinction being whether the estimate of the orbit is improved with each additional observation or at intervals, with all observations prior to the update processed at once.

Initial orbit determination is much the older subject. The first application of IOD dates to the discovery of the first of the minor planets, now called asteroids. On January 1, 1801, the first night of the Nineteenth Century, the Italian astronomer Piazzi of Palermo discovered Ceres, the largest asteroid (having a diameter of approximately 770 km). Piazzi observed Ceres for several weeks but then lost it. Shortly afterward Ceres approached conjunction and was thus too close to the sun to be seen. The problem of how to determine the likely position of Ceres as it emerged from conjunction became known to Gauss. Gauss determined an orbit for Ceres based on Piazzi's observations and predicted its position based on this orbit. Ceres was found again on January 1, 1802 within 0.5° (equal to the angle subtended by the full moon) of the position Gauss had predicted for it.

There is no question that Gauss developed a successful method of orbit determination from (angular position) observations. There is some controversy regarding precisely how he did it [10.1], as Gauss himself says in his *Theoria Motus* of 1809 that "··· scarcely any trace of resemblance remains between the method in which the orbit of Ceres was first computed, and the form given in this work." This ambiguity is reflected in the fact that today many different methods of IOD carry the appellation *Gauss' method*.

The characteristic these methods have in common is Gauss' observation that, since any two position vectors of the object determine the plane of the orbit, the position vector corresponding to any third observation must be capable of being expressed as a linear combination of the previous two. When this fact is coupled with a solution of the two-body problem of relative motion, in which position and velocity are expressed in a simple way as a function of the initial conditions, (i.e., in terms of the position and velocity at some epoch), a set of algebraic equations is derived which is sufficient in number) to allow the position (vectors) of the object to be determined at the three observation times. One can think of the method as a transformation of six independent pieces of information (e.g., the azimuth and elevation of the object with respect to the observer at three observation times), into the six (independent) components of the three position vectors of the object in the plane of the orbit. From these vectors another transformation yields the six elements of the orbit (see Sec. 3.3).

Another family of IOD methods is based on a method due to Laplace. The common feature of these methods is the use of the two-body equation of relative motion, rather than the solution for that motion. The acceleration of the object, which appears on one side of the equation of motion, is a function, among other things, of the angular position, angular velocity, and angular acceleration of the object, which may be observed or estimated from (again, a minimum of three) observations. A set of algebraic equations may then be solved for the position and velocity vectors of the object at one time and the methods of Sec. 3.3 again used to find the elements of the orbit.

As described earlier, both the Gauss and Laplace methods require three observations of the object's angular position, and are called *angles-only* IOD methods. IOD from three observations is of more theoretical than practical interest because more information is usually available. Even with more observations available, however, IOD may be accomplished with any three observations from the set, so that is still the fundamental technique. Additional observations raise additional questions (e.g., while any three allow the determination of the orbit in principle, different choices may yield systems of algebraic equations with significantly different degrees of difficulty of solution). This may manifest itself in reduced accuracy of the solution: three very closely spaced observations would seem intuitively to be bad, as they contain poorer quality information than more widely spaced observations (they clearly do not establish the orbit plane as well, for example), but widely spaced observations, as will be shown in the next section, can create problems for the convergence of the solution. Another practical difficulty is that, while it is ignored in the theory of IOD, observations in the real world are corrupted by noise and measurement errors, so it is not a priori obvious which observations to use and which (if any) to discard.

Observation types that were not available to Gauss and Laplace, such as radar-determined range and range-rate of an object, are common today. But again, they do not make these two-century-old methods obsolete; they simply reduce the number of unknowns being sought. There are however specific methods available which more efficiently incorporate the radar data, although they will not be described in detail here. The reader is referred to other texts [10.2] for a detailed discussion of these methods. One exception is the problem of determining an orbit from *two positions and times*, that is, when the observation includes range so that the position vectors at the observation times are known. This was already discussed in Chap. 4 as *Lambert's Problem*, the problem of finding the orbit that passes through two specified points in a given flight time, though the problem was approached from a mission design perspective and not as a problem in IOD.

10.2 Angles-Only Orbit Determination

Angles-only orbit determination has not been made obsolete by radar. For some objects, principally newly discovered asteroids, angular position data is all that is available. earth-orbiting objects will eventually be tracked by radar, but the type of observations of a satellite will depend on what facilities are available at the first tracking station to which it is visible.

Two methods of angles-only IOD will be described. As mentioned in the previous section, the kernel of one method is due to Laplace, the other to Gauss. Neither method will be described in its original form; there are more convenient ways to formulate the solution when a computer can be used to do the extensive calculations.

The first step in the use of either method is to convert the measurements of angular position, commonly either azimuth and elevation with respect to the tracking station or right ascension and declination, into a vector of direction cosines with respect to a space-fixed reference frame. It is convenient to locate the origin of this reference frame at the attracting center, as is shown in Fig. 10.1 for both the planetocentric and heliocentric cases. This transformation of a pair of angle measurements into a set of three direction cosines with respect to a space-fixed reference frame is straightforward but may involve several transformation matrices and requires knowledge of the date and time of the observation as well as, for the planetocentric case, the latitude and longitude of the tracking station. It will not be described here but may be found in several popular textbooks [10.3, 10.4]. The same set of three vectors may be used to describe the position of the object for both the planetocentric and heliocentric cases. In both Figs. 10.1a and 10.1b the vector $\mathbf{r}$ represents the absolute position of the object, ρ is the vector from the attracting center to the observer, and $\mathbf{R}$ is the position of the object with respect to the observer. We may write

$$\mathbf{r} = \mathbf{R} + \rho = R\hat{\mathbf{L}} + \rho \tag{10.1}$$

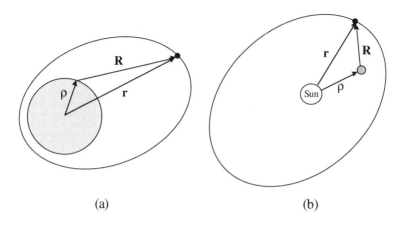

Fig. 10.1 Geometry for Planetocentric (a) and Heliocentric (b) Orbit Determination

where $\hat{\mathbf{L}}$ is the vector of topocentric direction cosines.

10.3 Laplacian Initial Orbit Determination

From Eq. (10.1) we may write the absolute acceleration of the object as

$$\ddot{\mathbf{r}} = \ddot{R}\hat{\mathbf{L}} + 2\dot{R}\dot{\hat{\mathbf{L}}} + R\ddot{\hat{\mathbf{L}}} + \ddot{\boldsymbol{\rho}} \qquad (10.2)$$

Combining (10.2) and the two-body equation of motion (1.24) yields

$$\frac{-GM\mathbf{r}}{r^3} = \ddot{R}\hat{\mathbf{L}} + 2\dot{R}\dot{\hat{\mathbf{L}}} + R\ddot{\hat{\mathbf{L}}} + \ddot{\boldsymbol{\rho}} \qquad (10.3)$$

R can be isolated by multiplying both sides of (10.3) with a vector that is normal to both $\hat{\mathbf{L}}$ and $\dot{\hat{\mathbf{L}}}$. One such vector is $\hat{\mathbf{L}} \times \dot{\hat{\mathbf{L}}}$, yielding

$$\frac{-GM\mathbf{r} \cdot (\hat{\mathbf{L}} \times \dot{\hat{\mathbf{L}}})}{r^3} = R\ddot{\hat{\mathbf{L}}} \cdot (\hat{\mathbf{L}} \times \dot{\hat{\mathbf{L}}}) + \ddot{\boldsymbol{\rho}} \cdot (\hat{\mathbf{L}} \times \dot{\hat{\mathbf{L}}}) \qquad (10.4)$$

or

$$R = \frac{-1}{\ddot{\mathbf{L}} \cdot (\hat{\mathbf{L}} \times \dot{\hat{\mathbf{L}}})} [\frac{GM}{r^3} \mathbf{r} \cdot (\hat{\mathbf{L}} \times \dot{\hat{\mathbf{L}}}) + \ddot{\boldsymbol{\rho}} \cdot (\hat{\mathbf{L}} \times \dot{\hat{\mathbf{L}}})] \qquad (10.5)$$

But

$$(\hat{\mathbf{L}} \times \dot{\hat{\mathbf{L}}}) \cdot \mathbf{r} = (\hat{\mathbf{L}} \times \dot{\hat{\mathbf{L}}}) \cdot (R\hat{\mathbf{L}} + \boldsymbol{\rho}) = (\hat{\mathbf{L}} \times \dot{\hat{\mathbf{L}}}) \cdot \boldsymbol{\rho}$$

since $\hat{\mathbf{L}} \times \dot{\hat{\mathbf{L}}}$ is normal to $\hat{\mathbf{L}}$.

Therefore

$$R = \frac{-1}{\ddot{\mathbf{L}} \cdot (\hat{\mathbf{L}} \times \dot{\hat{\mathbf{L}}})} [\frac{GM}{r^3} \boldsymbol{\rho} \cdot (\hat{\mathbf{L}} \times \dot{\hat{\mathbf{L}}}) + \ddot{\boldsymbol{\rho}} \cdot (\hat{\mathbf{L}} \times \dot{\hat{\mathbf{L}}})] \qquad (10.6)$$

This equation has two unknowns, r and R, assuming the direction cosine vector $\hat{\mathbf{L}}$ and its first derivative are known. A second equation involving r and R is easily generated using (10.1):

$$r^2 - [R^2 + \rho^2 + 2R(\hat{\mathbf{L}} \cdot \boldsymbol{\rho})] = 0 \qquad (10.7)$$

Equations (10.6) and (10.7) yield, by squaring both sides of (10.6) and substituting for R^2 from (10.7), a single polynomial equation of eighth degree in r.

In determining the coefficients of the polynomial, $\boldsymbol{\rho}$ and $\ddot{\boldsymbol{\rho}}$ can be determined analytically, $\hat{\mathbf{L}}$ is observed, but $\dot{\hat{\mathbf{L}}}$ and $\ddot{\hat{\mathbf{L}}}$ need somehow to be derived from the observations. Mean angular velocity may be obtained from any two observations; mean angular acceleration requires two estimates of velocity, thus a minimum of three observations is required to use this method. A better estimate of the first and second derivatives can be found, for example, by using Lagrange's interpolation formula [10.5],

$$\hat{\mathbf{L}}(t) = \hat{\mathbf{L}}(t_o) \frac{(t-t_1)(t-t_2) \cdots (t-t_m)}{(t_o-t_1)(t_o-t_2) \cdots (t_o-t_m)}$$

$$+ \hat{\mathbf{L}}(t_1) \frac{(t-t_1)(t-t_2) \cdots (t-t_m)}{(t_1-t_o)(t_1-t_2) \cdots (t_1-t_m)} + \cdots \qquad (10.8)$$

$$+ \hat{\mathbf{L}}(t_m) \frac{(t-t_1)(t-t_2) \cdots (t-t_{m-1})}{(t_m-t_o)(t_m-t_1) \cdots (t_m-t_{m-1})}$$

The first and second time derivatives of (10.8) yield (for the case of three observations),

$$\dot{\hat{\mathbf{L}}}(t) \approx \hat{\mathbf{L}}(t_o) \frac{(t-t_1)+(t-t_2)}{(t_o-t_1)(t_o-t_2)}$$

$$+ \hat{\mathbf{L}}(t_1) \frac{(t-t_o)+(t-t_2)}{(t_1-t_o)(t_1-t_2)} \qquad (10.9)$$

$$+ \hat{\mathbf{L}}(t_2) \frac{(t-t_o)+(t-t_1)}{(t_2-t_o)(t_2-t_1)}$$

and

$$\ddot{\hat{\mathbf{L}}}(t) \approx \hat{\mathbf{L}}(t_o) \frac{2}{(t_o-t_1)(t_o-t_2)}$$

$$+ \hat{\mathbf{L}}(t_1) \frac{2}{(t_1-t_o)(t_1-t_2)} \qquad (10.10)$$

$$+ \hat{\mathbf{L}}(t_2) \frac{2}{(t_2-t_o)(t_2-t_1)}$$

A very efficient method for the determination of the root of the eighth-degree polynomial is the Laguerre method of Sec. 2.5. The quantity n appearing in the Laguerre iteration formula (2.24) will now be 8. Convergence to the desired positive real root is assured if, as in its use in solving Kepler's equation, the absolute value of the quantity in the denominator of (2.24) is taken before the square root is found *and* if any negative real root found during the iteration process is rejected. This is accomplished by simply reversing the sign of any negative real root (estimate) encountered during the iteration process and continuing the iteration. The Laguerre method will then converge to a positive real root regardless of the initial guess for r.

Determining the orbit requires a knowledge of both $\mathbf{r}$ and $\dot{\mathbf{r}}$. The value of $\mathbf{r}$ is known when R is found, by substitution into Eq. (10.6) of the root r of the eighth-degree polynomial. From (10.1),

$$\dot{\mathbf{r}} = \dot{R}\hat{\mathbf{L}} + R\dot{\hat{\mathbf{L}}} + \dot{\boldsymbol{\rho}} \qquad (10.11)$$

The value of $\dot{R}$ is required, but from (10.3),

$$-\frac{GM}{r^3} \mathbf{r} \cdot (\hat{\mathbf{L}} \times \ddot{\hat{\mathbf{L}}}) = 2\dot{R}\dot{\hat{\mathbf{L}}} \cdot (\hat{\mathbf{L}} \times \ddot{\hat{\mathbf{L}}}) + \ddot{\boldsymbol{\rho}} \cdot (\hat{\mathbf{L}} \times \ddot{\hat{\mathbf{L}}}) \qquad (10.12)$$

or,

$$\dot{R} = \frac{-1}{2\dot{\hat{\mathbf{L}}} \cdot (\hat{\mathbf{L}} \times \ddot{\hat{\mathbf{L}}})} \, [\frac{GM}{r^3} \boldsymbol{\rho} \cdot (\hat{\mathbf{L}} \times \ddot{\hat{\mathbf{L}}}) + \ddot{\boldsymbol{\rho}} \cdot (\hat{\mathbf{L}} \times \ddot{\hat{\mathbf{L}}})] \qquad (10.13)$$

The orbit elements may now be determined using the methods of Sec. 3.3.

The previous analysis describes the Laplacian IOD in principle and has not emphasized the difficulties that are encountered in practice. The most significant of which is probably the solution of the eighth- degree polynomial in the unknown r. Unfortunately, there is ordinarily more than one positive real root of this equation; how does one determine which is correct? There are two simple approaches to resolving the problem, though they do not always work. The first approach is only available if there is some a priori knowledge of the orbit. Orbit element sets can be determined for each of the feasible values of r and then these sets compared on the basis of the partial knowledge of the orbit available (e.g., it may be known to be circular, or equatorial, or have a known period). If there is no a priori knowledge of the orbit, another technique is to use a different set of observations. Different roots of the polynomial equation will be obtained, but when orbit element sets are determined using each of these new solutions for r there should only be one element set, in principle, that will correspond exactly with one of those obtained using the previous set of observations. This test may not be decisive if the observations are closely spaced in time, that is, if they represent a viewing of only a "short-arc" of the orbit [10.6].

This is not the only difficulty encountered when one is trying to determine an orbit from a brief span of observations. It is obvious that a short arc of observations contains less "information" than an equivalent number of observations spread over a longer time. As mentioned previously, a short arc clearly determines the plane of the orbit much less accurately than a long one. This problem is evidenced in the Laplace method as an inaccurate determination of the first and second derivatives of $\hat{\mathbf{L}}$, for when only very small changes in $\hat{\mathbf{L}}$ are seen it is difficult to accurately estimate the velocity and acceleration of $\hat{\mathbf{L}}$. The problem is exacerbated in the real world since absolute errors in the tracking data, say in azimuth and elevation, are a larger fraction of the change in position of the satellite is a short-arc of data is collected as opposed to a longer one.

10.4 Gaussian Initial Orbit Determination

The kernel of all IOD methods referred to as Gauss' method is the observation that since the motion of the object occurs in a fixed plane, the position vector at any time must be capable of being expressed as a linear combination of the position vectors at any two other times:

$$\mathbf{r}_2 = c_1 \mathbf{r}_1 + c_3 \mathbf{r}_3 \qquad (10.14)$$

Using (10.1), Eq. (10.14) becomes

$$\mathbf{R}_2 - c_1 \mathbf{R}_1 - c_3 \mathbf{R}_3 = c_1 \boldsymbol{\rho}_1 + c_3 \boldsymbol{\rho}_3 - \boldsymbol{\rho}_2 \qquad (10.15)$$

but c_1 and c_3 are unknown. However, from (10.14),

$$c_1 = \frac{|\mathbf{r}_2 \times \mathbf{r}_3|}{|\mathbf{r}_1 \times \mathbf{r}_3|}$$

Orbit Determination

$$c_3 = \frac{|\mathbf{r}_2 \times \mathbf{r}_1|}{|\mathbf{r}_3 \times \mathbf{r}_1|} \quad (10.16)$$

One approach to solving for c_1 and c_3 is to use the f and g series of Sec. 2.4. Using the notation of Sec. 2.4, $\mathbf{r}_1$ and $\mathbf{r}_3$ may then be expressed in terms of $\mathbf{r}_2$ and $\mathbf{v}_2$,

$$\mathbf{r}_1 = f_1 \mathbf{r}_2 + g_1 \mathbf{v}_2 \; ; \; \mathbf{r}_3 = f_3 \mathbf{r}_2 + g_3 \mathbf{v}_2 \quad (10.17)$$

where

$$f_3 \approx 1 - \frac{(t_3 - t_2)^2}{2} h_2 \; ; \; g_3 \approx (t_3 - t_2) - \frac{(t_3 - t_2)^3}{6} h_2$$

$$f_1 \approx 1 - \frac{(t_1 - t_2)^2}{2} h_2 \; ; \; g_1 \approx (t_1 - t_2) - \frac{(t_1 - t_2)^3}{6} h_2 \quad (10.18)$$

Now let $\tau_1 \equiv t_3 - t_2$, $\tau_3 \equiv t_2 - t_1$, and $\tau_2 \equiv t_3 - t_1$, then

$$f_3 \approx 1 - \frac{\tau_1^2}{2} h_2 \; ; \; g_3 \approx \tau_1 (1 - \frac{\tau_1^2}{6}) h_2$$

$$f_1 \approx 1 - \frac{\tau_3^2}{2} h_2 \; ; \; g_1 \approx -\tau_3 (1 - \frac{\tau_3^2}{6}) h_2 \quad (10.19)$$

so that

$$c_1 = \frac{|\mathbf{r}_2 \times \mathbf{r}_3|}{|\mathbf{r}_1 \times \mathbf{r}_3|} = \frac{|\mathbf{r}_2 \times (f_3 \mathbf{r}_2 + g_3 \mathbf{v}_2)|}{|(f_1 \mathbf{r}_2 + g_1 \mathbf{v}_2) \times (f_3 \mathbf{r}_2 + g_3 \mathbf{v}_2)|}$$

$$= \frac{g_3}{f_1 g_3 - f_3 g_1}$$

$$\approx \frac{\tau_1 (1 - \frac{\tau_1^2}{6}) h_2}{\tau_1 (1 - \frac{\tau_1^2}{6}) h_2 - \frac{\tau_1 \tau_3^2}{2} h_2 - [-\tau_3 (1 - \frac{\tau_3^2}{6}) h_2)(1 - \frac{\tau_1^2}{2} h_2)]}$$

$$\approx \frac{\tau_1}{\tau_2} [1 + \frac{(\tau_2^2 - \tau_1^2)}{6} h_2] \quad (10.20)$$

One can similarly show that

$$c_3 \approx \frac{\tau_3}{\tau_2} [1 + \frac{(\tau_2^2 - \tau_3^2)}{6} h_2] \quad (10.21)$$

Equation (10.15) becomes:

$$R_2\hat{L}_2 - c_1 R_1 \hat{L}_1 - c_3 R_3 \hat{L}_3 = c_1 \rho_1 + c_3 \rho_3 - \rho_2 \quad (10.22)$$

where:

c_1 and c_3 are functions of r_2 and t_1, t_2, t_3.

$\hat{L}_1, \hat{L}_2, \hat{L}_3$ are observationally determined.

ρ_1, ρ_2, ρ_3 represent the position of the observer with respect to the sun (heliocentric case) or to the center of the earth (planetocentric case), and are known.

Equation (10.22) thus represents a set of three scalar equations in the four unknowns R_1, R_2, R_3, and r_2. One more independent equation,

$$r_2^2 = R_2^2 + \rho_2^2 + 2R_2 \hat{L}_2 \cdot \rho_2 \quad (10.23)$$

may be added to yield a system that may be solved.

The following method of solution is due to Escobal [10.2]. Equation (10.14) may be written as

$$c_1 \mathbf{r}_1 + c_2 \mathbf{r}_2 + c_3 \mathbf{r}_3 = 0 \quad (10.24)$$

where $c_2 = -1$, or, using (10.15),

$$c_1 R_1 \hat{L}_1 + c_2 R_2 \hat{L}_2 + c_3 R_3 \hat{L}_3 = -c_1 \rho_1 - c_2 \rho_2 - c_3 \rho_3 \equiv \mathbf{J} \quad (10.25)$$

or,

$$\begin{bmatrix} L_{1x} & L_{2x} & L_{3x} \\ L_{1y} & L_{2y} & L_{3y} \\ L_{1z} & L_{2z} & L_{3z} \end{bmatrix} \begin{bmatrix} c_1 R_1 \\ c_2 R_2 \\ c_3 R_3 \end{bmatrix} = \mathbf{J} = [L] \begin{bmatrix} c_1 R_1 \\ c_2 R_2 \\ c_3 R_3 \end{bmatrix} \quad (10.26)$$

Therefore,

$$\begin{bmatrix} c_1 R_1 \\ c_2 R_2 \\ c_3 R_3 \end{bmatrix} = [L]^{-1} \mathbf{J} \quad (10.27)$$

where

$$[L]^{-1} \equiv [A] = \begin{bmatrix} a_{11} & a_{12} & a_{13} \\ a_{21} & a_{22} & a_{23} \\ a_{31} & a_{32} & a_{33} \end{bmatrix} \quad (10.28)$$

Now, let

$$c_1 = \frac{\tau_1}{\tau_2} [1 + \frac{(\tau_2^2 - \tau_1^2)}{6} h_2] = A_1 + B_1 h_2$$

Orbit Determination

$$c_3 = \frac{\tau_3}{\tau_2}[1 + \frac{(\tau_2^2 - \tau_3^2)}{6} h_2] = A_3 + B_3 h_2 \tag{10.29}$$

and define

$$\mathbf{A}^T = [A_1 \ -1 \ A_3] \ ; \ \mathbf{B}^T = [B_1 \ 0 \ B_3]$$

$$\mathbf{X}^T = [\rho_{1x} \ \rho_{2x} \ \rho_{3x}] \ ; \ \mathbf{Y}^T = [\rho_{1y} \ \rho_{2y} \ \rho_{3y}] \tag{10.30}$$

$$\mathbf{Z}^T = [\rho_{1z} \ \rho_{2z} \ \rho_{3z}] \ , \ \text{where} \ \rho_{1x} = \boldsymbol{\rho}_i \cdot \hat{\boldsymbol{i}}, \ \text{etc.}$$

It can then be shown from (10.27) that

$$R_2 = A_2^* + B_2^* h_2 \tag{10.31}$$

where

$$A_2^* = a_{21} \mathbf{A} \cdot \mathbf{X} + a_{22} \mathbf{A} \cdot \mathbf{Y} + a_{23} \mathbf{A} \cdot \mathbf{Z}$$

$$B_2^* = a_{21} \mathbf{B} \cdot \mathbf{X} + a_{22} \mathbf{B} \cdot \mathbf{Y} + a_{23} \mathbf{B} \cdot \mathbf{Z} \tag{10.32}$$

Similarly,

$$R_1 = \frac{A_1^* + B_1^* h_2}{c_1} \ ; \ R_3 = \frac{A_3^* + B_3^* h_2}{c_3} \tag{10.33}$$

where

$$A_1^* = -(a_{11} \mathbf{A} \cdot \mathbf{X} + a_{12} \mathbf{A} \cdot \mathbf{Y} + a_{13} \mathbf{A} \cdot \mathbf{Z})$$

$$B_1^* = -(a_{11} \mathbf{B} \cdot \mathbf{X} + a_{12} \mathbf{B} \cdot \mathbf{Y} + a_{13} \mathbf{B} \cdot \mathbf{Z})$$

$$A_3^* = -(a_{31} \mathbf{A} \cdot \mathbf{X} + a_{32} \mathbf{A} \cdot \mathbf{Y} + a_{33} \mathbf{A} \cdot \mathbf{Z})$$

$$B_3^* = -(a_{31} \mathbf{B} \cdot \mathbf{X} + a_{32} \mathbf{B} \cdot \mathbf{Y} + a_{33} \mathbf{B} \cdot \mathbf{Z}) \tag{10.34}$$

However, $h_2 = \mu/r_2^3$ is still unknown. Substituting from (10.31) into (10.23) yields,

$$r_2^2 = \rho_2^2 + (A_2^* + B_2^* \frac{\mu}{r_2^3})^2 + 2(\hat{\mathbf{L}}_2 \cdot \boldsymbol{\rho}_2)(A_2^* + B_2^* \frac{\mu}{r_2^3}) \tag{10.35}$$

or,

$$r_2^8 - r_2^6[\rho_2^2 + A_2^{*2} + 2(\hat{\mathbf{L}}_2 \cdot \boldsymbol{\rho}_2)A_2^*]$$

$$-r_2^3[2A_2^*B_2^*\mu + 2(\hat{\mathbf{L}}_2 \cdot \boldsymbol{\rho}_2)B_2^*\mu] - B_2^{*2}\mu^2 = 0 \tag{10.36}$$

We see that just as in Laplace's method the real root(s) of an eighth-degree polynomial must be found.

The solution is guaranteed to be inexact because we have used truncated forms of the series expansions for f and g in the determination of the quantities c_1 and c_3 [see Eq. (10.18)]. Higher-order terms in these series require a knowledge of the velocity vector (at time t_2). The velocity cannot be determined until the orbit is known, which it is in principle (though only approximately) after this "first-pass" solution for R_1, R_2, R_3, and r_2 and hence $\mathbf{r}_1$, $\mathbf{r}_2$, and $\mathbf{r}_3$ is obtained. The "obvious" solution is to use the estimate of the velocity at the intermediate time t_2 to calculate higher-order terms in the f and g series, solve the resulting equations again, and determine the velocity at t_2 again. This process could be continued until no significant change in the orbit [or in $\mathbf{v}(t_2)$] is observed.

The Gauss method, as described earlier, has been termed "not acceptable" [10.1] since it may not converge to a solution because of the small radius of convergence of the f and g series, so that one does not know a priori if the spacing of the observations in time is too great for successful IOD. Another researcher [10.7] says that this criticism is "completely unwarranted" since one may use instead the f and g *functions* (of Sec. 2.4) that allow f and g to be determined exactly. This is not strictly true, since to use the f and g functions, of Eq. (2.26) or in the universal form (2.38), the velocity must be known. In the process described earlier, the velocity of the object is not known until after a "first-pass" through the orbit determination algorithm. The f and g series must thus be used for at least the first approximation to the orbit. For all subsequent iterations there is no good reason to use the f and g *series*, even if additional terms in the series are calculated, contrary to what was said in the previous paragraph. Instead one should use the f and g *functions*, as these will yield "exact" values for c_1 and c_3. This should benefit the convergence of the method but will not eliminate the problem. It is still possible for the orbit and the spacing of the observations to be such that the first approximation to the orbit, found of necessity using the f and g series, may be so poor that the method will not converge to a solution.

10.5 Orbit Determination from Two Position Vectors

This subject will not be discussed here in any detail because it has already been described in Chap. 4. However, there it was referred to as Lambert's problem and presented as a problem in orbit transfer, not orbit determination. The principle is the same in both cases; we wish to determine what orbit will connect two points in space with a specified time of flight.

Orbit Determination 181

To begin the orbit determination process, the absolute position of the object at two times is needed. This will be determined from Eq. (10.1),

$$\mathbf{r} = \mathbf{R} + \boldsymbol{\rho} \tag{10.1}$$

where **R** is now a known vector, obtained using radar range information. One may then proceed as described in Sec. 4.6 to obtain the semimajor axis of the orbit and the terminal velocity vectors, which are the velocity vectors at the observation times. From either set of position and velocity vectors the orbit elements may be determined using the method of Sec. 3.3.

10.6 Differential Correction

While the preceding analysis has described the problem of determining an orbit using the minimum number of observations, in practice there are usually more observations available both at the time of initial orbit determination and, especially for earth-orbiting objects, at regular intervals thereafter. If the observations were all perfect and if the orbit was not changing with time, the orbit could be determined by any subset of three angle measurements or two position-vector measurements with the same result; however, this is not the case. Angle and range measurements are corrupted by small systematic and random errors, some tracking stations have smaller measurement errors than do others, and the orbit is changing with time due to such things as atmospheric drag, the gravitational attraction of the nonspherical and nonuniform earth, and the attraction of the sun and moon, as discussed in Chap. 9. How can one determine the best estimate of the state (the orbital elements) of the system under these circumstances? This is the subject of estimation theory, a science in its own right, from which only a few concepts and results will be extracted and presented in this section. The interested reader is referred to two of the many texts [10.8, 10.9] on estimation theory.

Since the IOD process does not make any use of a priori knowledge of the orbit (which may span a long history and which may have been accumulated at great expense), the most reasonable course is to use subsequent measurements to improve the initial estimate of the orbit and "maintain" the elements of the orbit which change with time due to perturbations. This is differential correction, so named because the DC process attempts to find (or estimate) only two presumably small quantities: the error in the IOD result and the change in the orbit due to the perturbations. In determining the latter, the process is aided by having a theory to govern some of the effects of the perturbations (e.g., the first-order effects of earth oblateness and atmospheric drag derived in Chap. 9). The process is also aided by having estimates of the errors to be expected from each tracking station, determined over many years of operation.

There are two basic types of differential correction, — *batch* and *sequential*, — depending on whether the measurement data is processed simultaneously with the measurements being stored until the processing time (the *update*), or sequentially with the update occurring with each measurement. One example of each type of process or *filter* will be described. In both cases we will assume that measurements are made at discrete times.

Assume that the system governing equations are:

$$\dot{\mathbf{x}} = \mathbf{f}[\mathbf{x}(t), t] + \mathbf{w}(t) \qquad (10.37)$$

where $\mathbf{x}$ is the system state vector (e.g., the six orbital elements augmented by the ballistic coefficient), a measure of the drag experienced by the satellite. Then the first six elements of the column vector $\mathbf{f}[\mathbf{x}(t), t]$ are the rates of change of the orbital elements due to perturbations, described by the Lagrange equations discussed in Chap. 9. The function $\mathbf{w}(t)$ represents the random error, caused, for example, by unmodeled dynamics. Usually $\mathbf{w}(t)$ is represented as a Gaussian or normal process with *zero mean* and *covariance matrix* $Q(t)$. By zero mean it is meant that $E[w_i(t)] = 0$, $i = 1, 2, ..., 7$; the covariance matrix Q is the matrix

$$Q = \begin{bmatrix} E[w_1^2] & E[w_1 w_2] & \cdot & E[w_1 w_7] \\ E[w_2 w_1] & E[w_2^2] & \cdot & \cdot \\ \cdot & \cdot & \cdot & \cdot \\ \cdot & \cdot & \cdot & \cdot \\ E[w_7 w_1] & E[w_7 w_2] & \cdot & E[w_7^2] \end{bmatrix}$$

where $E[\ ]$ stands for the *mean* or *expected value* of its argument. The mean value of a continuous random variable $y(t)$ is given by

$$E[y] = \int_{-\infty}^{\infty} y\, p(y)\, dy$$

where $p(y)$ is the *probability density function* for the process $y(t)$. For a particular value of y, designated by Y, the value of $p(Y)$ represents how likely it is for the random variable y to assume a value near Y. Specifically, $p(Y)\, dY$, which may be interpreted as the area under the probability density function $p(y)$ between $y = Y$ and $y = Y + dY$, is the probability that y will assume a value in the range $Y \le y \le Y + dY$. With regard to the covariance matrix, the usual simplifying assumption is that $E[w_i w_j] = 0$ if $i \ne j$ (i.e., the process errors are uncorrelated).

The measurement vector is $\mathbf{z}(t)$ and is usually a two-element vector consisting of azimuth and elevation or right ascension and declination.

Orbit Determination

$$\mathbf{z}(t) = \mathbf{h}[\mathbf{x}(t), t] + \mathbf{v}(t) \tag{10.38}$$

The vector $\mathbf{v}(t)$ is the random error and will also be represented as a zero-mean, normal process with covariance matrix $R(t)$. Typical values for the mean error on measuring azimuth and elevation are $0.002 - 0.005°$ ($3.5 \cdot 10^{-5} - 8.7 \cdot 10^{-5}$ radians.) A typical value for the R matrix might then be (in radians^2):

$$R = \begin{bmatrix} 1.22 \cdot 10^{-9} & 0 \\ 0 & 1.22 \cdot 10^{-9} \end{bmatrix}$$

The measurements are a nonlinear function $\mathbf{h}[\mathbf{x}(t), t]$ of the states. Define

$$H[\mathbf{x}(t), t] = \frac{\partial \mathbf{h}[\mathbf{x}(t), t]}{\partial \mathbf{x}(t_o)} \tag{10.39}$$

The matrix H represents the sensitivity of the measurement vector to changes in the state vector at epoch t_o (the time of the IOD). The *weighted-least-squares* estimate of the state may then be found using the following process:

1. Select a batch size, m for processing the data.
2. Read in m sets (here, pairs) of observations (measurements).
3. Propagate the state from epoch to each observation time t_i, using

$$\dot{\mathbf{x}}(t) = \mathbf{f}[\mathbf{x}(t), t] \tag{10.40}$$

   with initial condition $\mathbf{x}(t_o) =$ current best estimate of the state $\equiv \overline{\mathbf{x}}$. This is easily done using a computer program for integration of systems of ODEs [10.10].
   Calculate the *predicted observations* $\overline{\mathbf{z}}(t_i) = \mathbf{h}[\mathbf{x}(t_i), t_i]$.
   Calculate the sensitivities $H[\mathbf{x}(t_i), t_i]$.
   Combine the $\overline{\mathbf{z}}(t_i)$ to form a $2m \times 1$ column vector of predicted observations.
   Combine the $\mathbf{z}(t_i)$ to form a $2m \times 1$ column vector of *actual* observations.
   Combine the $H[\mathbf{x}(t_i), t_i]$ to form a $2m \times 7$ matrix.
   Combine the m 2×2 measurement error covariance matrices into one $2m \times 2m$ matrix R.

4. Calculate $\hat{\mathbf{x}}$, the *best estimate of the state*, using

$$\hat{\mathbf{x}} = \overline{\mathbf{x}} + (H^T R^{-1} H)^{-1} H^T R^{-1} (\mathbf{z} - \overline{\mathbf{z}}) \tag{10.41}$$

   where $\overline{\mathbf{x}}$ is the previous best estimate of the state at epoch.

5. The vector $\hat{\mathbf{x}}$ is the new best estimate of the state at epoch.

   Now go to step 3, using $\hat{\mathbf{x}}$ as the new initial condition for the integration of Eq. (10.40). Continue steps 3 – 5 until $|\hat{\mathbf{x}} - \bar{\mathbf{x}}|$ becomes small.

The estimate is called "weighted" because, through the matrix R, the uncertainty in each measurement, reflecting the expected error from each tracking station, is accounted for. Only R^{-1} appears in the update equation (10.41), so that more uncertain measurements are weighted less significantly. Note that the "process error" $\mathbf{w}(t)$ is not explicitly accounted for by the weighted-least-squares filter. Process noise can be included.

There are many sequential filters, but the one that will be described here is the *extended Kalman filter* [10.9]. Introduce the estimation error covariance matrix $P(t)$,

$$P(t) = E[(\hat{\mathbf{x}}(t) - \mathbf{x}(t))(\hat{\mathbf{x}}(t) - \mathbf{x}(t))^T] \tag{10.42}$$

where $\hat{\mathbf{x}}$ continues to represent the best estimate of the true state $\mathbf{x}$. By analogy with (10.39), introduce

$$F[\mathbf{x}(t), t] = \frac{\partial \mathbf{f}[\mathbf{x}(t), t]}{\partial \mathbf{x}(t)} \tag{10.43}$$

where the matrix F represents the sensitivity of the system dynamics (the Lagrange variational equations) to changes in the state vector. Let $\hat{\mathbf{x}}(t_o)$ represent the best estimate of the state at the epoch time. The extended Kalman filter best estimate of the state is then given by

$$\hat{\mathbf{x}}_k(t) = \hat{\mathbf{x}}_k(-) + K_k[\mathbf{z}_k - \mathbf{h}(\hat{\mathbf{x}}_k(-))] \tag{10.44}$$

where the "gain" K_k is given by

$$K_k = P_k(-)H_k^T[\hat{\mathbf{x}}_k(-)]\Big[H_k[\hat{\mathbf{x}}_k(-)]P_k(-)H_k^T[\hat{\mathbf{x}}_k(-)] + R_k\Big]^{-1} \tag{10.45}$$

and the error covariance update is given by

$$P_k(+) = \Big[I - K_k H_k[\hat{\mathbf{x}}_k(-)]\Big] P_k(-) \tag{10.46}$$

where:

I is the identity matrix
$\hat{\mathbf{x}}_k(-)$ is $\hat{\mathbf{x}}(t_k)$ *before* processing of the measurements from $t = t_k$
$\hat{\mathbf{x}}_k(+)$ is $\hat{\mathbf{x}}(t_k)$ *after* processing of the measurements from $t = t_k$
$P_k(-)$ is $P(t_k)$ *before* processing of the measurements from $t = t_k$
$P_k(+)$ is $P(t_k)$ *after* processing of the measurements from $t = t_k$

and

$$H_k[\hat{\mathbf{x}}_k(-)] = \left[\frac{\partial \mathbf{h}[\mathbf{x}(t_k), t_k]}{\partial \mathbf{x}(t_k)}\right]_{\mathbf{x} = \hat{\mathbf{x}}_k(-)}$$

Between updates, the state vector and the estimation error covariance matrix are propagated forward with

$$\dot{\hat{\mathbf{x}}}(t) = \mathbf{f}[\hat{\mathbf{x}}(t), t] \ ; \ \hat{\mathbf{x}}(t_k) = \hat{\mathbf{x}}_k(+) \tag{10.47}$$

$$\dot{P}(t) = F[\hat{\mathbf{x}}(t), t] P(t) + P^T(t) F^T[\hat{\mathbf{x}}(t), t] + Q \ ; \ P(t_k) = P_k(+)$$

Note that:

1. With the sequential filter, the process error $\mathbf{w}(t)$ is incorporated through the appearance of $Q(t)$ in the forward propagation of the error covariance in Eq. (10.47).

2. The computation of the updated estimate using Eqs. (10.44) and (10.46) involves only the *current* measurement and error covariance matrix. Thus the updates can be done in real time with Eq. (10.47) numerically integrated between measurements.

3. If less-exact results would suffice, one can precompute and store the gains K_k by integrating the differential equation (10.47) forward *without* measurement updates using $\hat{\mathbf{x}}(t_o)$ and $P(t_o)$ as the initial conditions.

References

10.1 Taff, L. G., *Celestial Mechanics*, John Wiley & Sons, New York (1985)

10.2 Escobal, P. R., *Methods of Orbit Determination*, John Wiley & Sons, New York (1965), also reprint with corrections, Krieger Publishing Co., Malabar, FL (1976)

10.3 Roy, A. E., *Orbital Motion*, Adam Hilger, Ltd., Bristol, 2nd ed. (1982)

10.4 Bate, R. R., Mueller, D. D., and White, J. E., *Fundamentals of Astrodynamics*, Dover Publications, New York (1971)

10.5 Weeg, G. P., and Reed, G. B., Introduction to Numerical Analysis, Ginn & Co., Waltham, MA (1966)

10.6 Kaya, D. A., and Snow, D. E., "Short Arc Initial Orbit Determination Using Angles-Only Spaced-Based Observations," Paper AAS 91-358,

AAS/AIAA Astrodynamics Specialist Conference, Durango, CO (1991)

10.7 Marsden, B. G., "Initial Orbit Determination: The Pragmatists Point of View," *Astronomical Journal*, **90**, 8, 1541–1547 (1985)

10.8 Bryson, A. E. and Ho, Y. C., *Applied Optimal Control*, Hemisphere Publishing, New York (1975)

10.9 Gelb, A. (ed.) *Applied Optimal Estimation*, M.I.T. Press, Cambridge, MA (1974)

10.10 Shampine, L. F. and Gordon, M. K., *Computer Solution of Ordinary Differential Equations*, W. H. Freeman, San Francisco (1975)

Problems

10.1 a) Show that, as an alternative to using Lagrange's interpolation formula to provide estimates of $\dot{\hat{L}}$ [Eq. (10.9)] and $\ddot{\hat{L}}$ [Eq.(10.10)], one may use a truncated Taylor series. In particular, show that three observations $\hat{L}(t_1), \hat{L}(t_2), \hat{L}(t_3)$ provide sufficient information to solve for $\dot{\hat{L}}$ and $\ddot{\hat{L}}$ at intermediate observation time t_2. Solve explicitly for $\dot{\hat{L}}(t_2)$ and $\ddot{\hat{L}}(t_2)$.

b) Show that a fourth observation allows a more accurate determination of $\dot{\hat{L}}(t_2)$ and $\ddot{\hat{L}}(t_2)$ since one more term in the Taylor series of (a) may be included.

10.2 Derive relations (10.31) and (10.33).

10.3 a) Use the Laguerre root finding method to determine the root of the equation [cf. (10.36)]:

$$r_2^8 - .9945225\, r_2^6 - 3.370698\, r_2^3 - 365.7847 = 0$$

which results from applying Gauss' method to the determination of the orbit of an earth orbiting spacecraft.

Note: The orbit has a semimajor axis of 2.9 R_{earth} and an eccentricity of 0.3. This yields bounds on reasonable first approximations (starting values) for the iteration process.

b) Determine the possible values of the true anomaly of the orbit at the observation time.

c) Use the Laguerre root finding method to determine the root of the equation:

$$r_2^8 - .9779058\, r_2^6 - 18.5389\, r_2^3 - 607.8789 = 0$$

Use the very poor starting approximation $r_2 = 0.1$ (which would place the spacecraft near the center of the earth.) This guess should yield convergence to a (physically unreasonable) negative root. Use the method described in the text, of changing the sign of any negative intermediate value obtained during the iteration, to obtain the desired positive root of the equation from the same starting value.

10.4 a) Determine the definition of the expected value of a discrete random variable (e.g., the result of the roll of a die) using the definition in the text of the expected value of a continuous random variable.
b) What is the expected value of the outcome of a roll of a die?

Appendix 1
Astronomical Constants

The Sun

Mass = $1.989 \cdot 10^{30}$ kg

Radius = $6.9599 \cdot 10^5$ km

$\mu_{sun} = Gm_{sun} = 1.327 \cdot 10^{11}$ km³/s²

The Earth

Mass = $5.974 \cdot 10^{24}$ kg

Radius = $6.37812 \cdot 10^3$ km

$\mu_{earth} = Gm_{earth} = 3.986 \cdot 10^5$ km³/s²

Mean distance from sun = 1 au = $1.495978 \cdot 10^8$ km

The Moon

Mass = $7.3483 \cdot 10^{22}$ kg

Radius = $1.738 \cdot 10^3$ km

$\mu_{moon} = Gm_{moon} = 4.903 \cdot 10^3$ km³/s²

Mean distance from earth = $3.844 \cdot 10^5$ km

Orbit eccentricity = 0.0549

Orbit inclination (to ecliptic) = 5° 09′

Appendix 2
Physical Characteristics of the Planets

Planet	Equatorial Radius (units of R_{earth})	Mass (units of M_{earth})	Sidereal Rotation Period	Inclination of Equator to Orbit Plane
Mercury	0.382	0.0553	58d 16h	≈ 2°
Venus	0.949	0.8149	243d (retro)	177° 18'
Earth	1.000	1.000	23h 56m 04s	23° 27'
Mars	0.532	0.1074	24h 37m 23s	25° 11'
Jupiter	11.209	317.938	9h 50m	3° 07'
Saturn	9.49	95.181	10h 14m	26° 44'
Uranus	4.007	14.531	17h 54m	97° 52'
Neptune	3.83	17.135	19h 12m	29° 36'
Pluto	0.18	0.0022	6d 9h 18m	122° 46'

Appendix 3

Elements of the Planetary Orbits

Planet	Semimajor Axis (in au)	Eccentricity	Inclination to Ecliptic Plane	Sidereal Period
Mercury	0.3871	0.2056	7° 00′	87.969d
Venus	0.7233	0.0068	3° 24′	224.701d
Earth	1.000	0.0167	0° 00′	365.256d
Mars	1.5237	0.0934	1° 51′	1y 321.73d
Jupiter	5.2028	0.0483	1° 19′	11y 314.84d
Saturn	9.5388	0.0560	2° 30′	29y 167d
Uranus	19.1914	0.0461	0° 46′	84y 7.4d
Neptune	30.0611	0.0097	1° 47′	164y 280.3d
Pluto	39.5294	0.2482	17° 09′	247y 249d

Index

Aiming radius, 18, 61: *see also* Hyperbolic orbit
Angles-only orbit determination, 171
Angular momentum, 5
 specific, 15
Aphelion, 4, 5
Apoapse, 16
Areal velocity, 16
Argument of latitude at epoch, 48
Argument of periapse, 46
Ariane IV, 91
Ascending node, 47, 52
Astrodynamics, 3
Astronomical constants, 188
Astronomical unit, 45
Atmospheric
 density, 164
 drag, 157, 164, 168

Ballistic coefficient, 164
Battin-Vaughan algorithm, 74, 78
Bi-elliptic transfer, 108-13
Bi-parabolic transfer, 108-12, 119
Brahe T, 3

Canonical units, 23, 44, 45, 60, 74
Celestial latitude, 50
Celestial longitude, 50
Celestial mechanics, 3
Celestial sphere, 50
Center of mass, 3, 7
Ceres, 170
Characteristic velocity, 84, 89
Chord, 69
Circular velocity, 21
Colatitude, 165
Conic section, 14, 21, 26: *see also* Elliptic orbit, Hyperbolic orbit, Parabolic orbit
Conjunction, 114, 115

Covariance matrix, 182, 184
CW equations, 142-49: *see also* Hill-Clohessy-Wiltshire equations
Cylindrical coordinates, 146

DC, 170
Declination, 50, 60
Delta-vee, 84
Differential correction, 180-83
 Batch, 182
 Sequential, 182
Dirac delta function, 99
Directrix, 14
Disturbing function, 122
Drag coefficient, 164

Eccentric anomaly, 27, 45, 71
 solution for, 30-32
Eccentricity, 14
 of planetary orbits, 190
 of transfer orbit, 78, 79
 minimum, 67
 vector, 13, 52, 66
Ecliptic plane, 47
Elliptic orbit, 14, 26, 17-19, 56, 57
 minimum-energy, 66, 71
EMOS, 45
Energy
 constant, 18
 kinetic, 8, 18, 19, 131
 potential, 8, 18, 19, 131
 potential, of thin shell, 11, 23
 potential, of sphere, 11, 23
 specific, 19
Epoch, 36, 46, 47
Equatorial
 plane, 47, 52
 radius, 189
Equinoctial elements, 60

Escape
 velocity, 20
 hyperbola, 128
Estimation error, 184
Euler angles, 53, 160
Euler's equation, 74
Expected value, 182, 187
Extended Kalman filter, 184

f and g functions and series, 32-36, 177,180
Filter, 182
First point in Aries, 46, 47, 50
Flight path angle, 17, 22, 24, 53
Focus, 14
Force
 conservative, 63
 drag, 168
 external, 83, 87, 88
 gravitational, 5, 6
 internal, 83
 oblateness, 166
Fundamental ellipse, 67

Galileo, 3
Gauss K, 34, 170
Gauss'
 perturbation equations, 163
 method, 170, 176-80
Geocentric-equatorial frame, 50
Gooding procedure, 74, 78
Gravity, 5
 -assist, 87, 120, 129-34
 constant, 12, 133
 gradient matrix, 141, 154
 loss, 88
 universal constant of, 6
GEO, 103, 119
GTO, 96
Gyroscopic damping, 144

Harmonic oscillator, 25
Hill-Clohessy-Wiltshire equations, 142-149
Hohmann W, 104
Hohmann transfer, 103-108, 116, 126, 134, 152
Hyperbolic orbit, 14, 17-19, 56, 59
 aiming radius, 18
 turn angle, 18, 132
Hyperbolic excess velocity, 21

Impulsive thrust approximation, 99-102
Inclination, 46, 52
 of the planetary orbits, 190
Inertial reference frame, 6
Inferior planet, 114, 115
Initial phase angle, 116
Integral of the motion, 9
Interception, 62, 114-17, 140
Inverse-square law, 5: *see* Gravity
Invariable plane, 8
IOD, 170

Kepler J, 3, 26
Kepler's
 equation, 27, 29-30, 68
 equation, bounds on solution, 32
 equation, in universal variables, 37, 43
 equation, in universal variables, bounds on solution, 39, 40
 laws, 3
 first law, 4
 second law, 5, 6, 16, 28
 third law, 5, 17
Kinetic energy, 8, 18, 131: *see also* Energy

Lagrange J L, 68
Lagrange
 multiplier, 92
 interpolation formula, 174
Laguerre algorithm, 38-41, 175,
Lambert J H, 67
Lambert's
 equation, 70
 equation, solution to, 70-75
 problem, 67, 101, 150, 172, 180
 theorem, 67, 68
Laplace P S, 8, 121, 171
Laplacian IOD, 173-76
Legendre polynomial functions, 165
LEO, 42, 89, 103

Index 193

Line of nodes, 46, 52
Linearized orbit theory, 139-44
Local vertical coordinates, 142
Longitude of the ascending node, 47, 53
Longitude of periapse, 48

Mass
  flow rate, 81
  payload, 88
  planetary, 189
  propellant, 88
  ratio, 84, 88
  structural, 88
Mean anomaly, 27
Mean distance, 5
Mean motion, 27
Measurement vector, 182
Minimum-energy ellipse, 66, 71
Momentum
  angular, 5, 16, 156
  linear, 7, 82

N-body problem, 6, 9, 22
Newton I, 5, 11, 12
Newton's
  law of gravity, 6
  method, 30, 38
  second law, 6, 81, 82
Nodal vector, 52
Noise, 171

Oblateness, 164-68
Obliquity
  of the ecliptic, 51
Observations, 183
Opposition, 114, 115
Orbit
  cranking, 112
  elements, 46-54, 60
  period, 16
  plane, 13
  rectilinear, 18, 71
  reference, 143
  target, 146
  transfer, 62-67, 99-114

Parabolic orbit, 14, 17-19, 56, 58
Parameter, 15: see Semi-latus rectum
Patched conic method, 120, 124-28
Payload ratio, 89
Periapse, 15
Perigee, 15
Perilune, 15
Perihelion, 4, 5
Period, 16
  of the planetary orbits, 190
  sidereal, 119, 189
  synodic, 114, 116
Perturbation, 10, 123, 155
  equations, 155-63: see also Variational equations, Atmospheric drag, Oblateness, Third-body effect
Piazzi, 170
Plane change angle, 112, 114
Planetary flyby, 120, 129-37: see also Gravity-assist
Position
  vs. time, on elliptic orbit, 26-30
  vs. time, on hyperbolic and parabolic orbits, 36-41: see also Universal variables
Potential energy, 8: see also Energy
Precession, 47, 48
Probability density function, 182
Process error, 184

Quadrature, 9

Radar, 49, 171
Random error, 182, 183: see also Noise
Rectilinear orbit, 18, 71
Regression of node, 166
Relative motion, 15, 140
Relative position vector, 139
Rendezvous, 62, 114-17, 150-53
Right ascension, 50, 60
Rocket
  engine, 81
  equation, 81-88
Rotation period, 189

Secular term, 148, 166

Semi-latus rectum, 15
Semimajor axis, 5
  of planetary orbits, 190
Semiminor axis, 15
Semiperimeter, 66
Signum function, 74
Singular elements, 49
SOI, 121
Space triangle, 62
Specific angular momentum, 15: *see* momentum
Specific impulse, 85
Sphere of influence, 120-24, 133
  radii, for planets, 125
Staging, 88-92
  optimal, 92-96
State transition matrix, 149
State vector, 149
Structural coefficient, 89
Successive approximations, method of, 30: *see also* Newton's method, Laguerre algorithm, Gooding procedure, Battin-Vaughan algorithm
Sun-synchronous orbit, 168
Superior planet, 114, 115
Swingby trajectory, 129: *see also* Planetary flyby trajectory
Synodic period, 114, 116
Syzygy, 114

Terminal radius ratio, 108
Terminal velocity vectors, 75-79
Third body effect, 122
Thrust, 81, 83
  high and low, 86
  impulsive, 99-102
Time
  canonical units, 74
  epoch, 36
  of periapse passage, 27
  parabolic transfer, 72, 78

Transcendental equation, 29
Transcendental functions, 37
Transfer
  angle, 62, 77, 115
  bi-elliptic, 108-113
  bi-parabolic, 108-12, 119
  Hohmann, 103-108, 116, 126, 134, 152
  minimum-energy, 102
  minimum-fuel, 102, 106, 110, 111: *see also* Hohmann transfer
  multiple revolution, 79
  orbits, 62-67, 99-114
  parabolic, 72, 78, 80
  time, 70, 71, 75, 104, 109
  two-impulse, 102, 103
True anomaly, 14, 28: *see also* Orbit elements
True longitude at epoch, 48
Turn angle, 18, 61, 132: *see also* Hyperbolic orbit
Two-body problem, 9, 12-15

Unit impulse, 99
Universal variables, 36-42

Vacant focus, 64
Variational equations, 159
Velocity
  burnout, 97
  characteristic, 84, 89
  circular, 21, 23
  effective exhaust, 81
  escape, 20
  hodograph, 54-59
  hyperbolic excess, 21, 124, 126
  vectors, terminal, 75-78
Vernal equinox, 47, 48
Vis-Viva equation, 20, 51

Weighted averaging, 49, 184
Weighted-least-squares, 183